Contemporary Social
and
Sociological Theory

Having a Thought

Formerly, one could tell simply by looking at a person that he wanted to think . . . that he now wished to become wiser and prepared himself for a thought: he set his face as for prayer and stopped walking; yes, one even stood still for hours in the middle of the road when the thought arrived—on one leg or two legs. That seemed to be required by the dignity of the matter.

(Nietzsche, 1974, p. 81)

D id you read the quote? If not, please do so. And let what Nietzsche says get inside of you. Have you ever had a thought? Of course you think quite a bit. But have you ever had a thought in the way Nietzsche is describing it? This kind of thought is an event. It requires or perhaps captures the entire person. It's demanding and inspiring. Notice that Nietzsche says "formerly." He's referring to a time before the seduction of modern busyness. When life was slower, it was easier to have a thought, to be captured by an idea when you least expected it. Most of us today are too busy to be taken over by an idea—but we can still quite deliberately have a thought. And that's what this book is about. I'm inviting you to have a thought or two with me. It will require time and effort. But once you've had a thought, you'll never be the same again.

The Organization of the Book

Every book tells a story. That's probably pretty obvious when we are talking about novels, but theory books tell stories too. The most apparent way they do is through the text: In each chapter of this book, we review someone's theory, and each of those theories tells a story about society. But theory books tell stories in much more subtle ways as well. Sometimes these more subtle ways are the most powerful. For

We will also see shifts in the way the individual and his or her political involvement are seen. We begin with the interactionist understanding of the self, which is the epitome of what a democratic citizen ought to be: free and responsible for his or her own choices. But in Chapter 12, Giddens tells us that societies in late modernity no longer practice emancipatory politics; instead, we have lifestyle politics, where the central issues are choice and the reflexive project of the self rather than political freedoms and rights. And we end the book considering identity politics—a very different sort of political action.

Our journey is filled with amazing questions and equally striking answers: What is society? What (rather than who) are you? Where do you fit in society? How do you fit in society? Are there global structures and a global system? Do we now live in a symbolic space where the local and global collide? How are economic shifts and the proliferation of mass media and advertising images influencing you, society, politics, and the world order? Is the social world empirical, as Herbert Blumer claims, or is it textual, as Jacques Derrida insists? Can we have a social science, or is the idea of the social sciences simply an effect of historical and political conditions, as Michel Foucault argues? Is there a generalized other, or does our very consciousness change by gender, as Dorothy Smith tells us? I hope that this book is an invitation to think. Let this book challenge you to find questions rather than answers.

One more interesting thread for us to consider before we embark on our trip: What about the body? In thinking about it, I'm going to leave you with two quotes, one from the person we begin this book with—George Herbert Mead—and one from the person we end the book with—Judith Butler:

It may be necessary again to utter a warning . . . against the assumption that the body of the individual as a perceptual object provides a center to which experiences may be attached, thus creating a private and psychical field that has in it the germ of representation and so of reflection. (Mead, 1934, p. 357)

[There] are certain constructions of the body constitutive in this sense: that we could not operate without them, that without them there would be no "I," no "we." (Butler, 1993, p. xi)

How to Use This Book

There are a few unique features in this book. First, the book is written in a conversational tone. I try to speak in academic jargon as little as possible. I use personal examples and try to convey my experience with theory—and I actually get pretty excited and caught up in this stuff. Second, I don't believe in any of these theories. That might sound a little odd, but I think it's a good thing. As much as possible, I try to let the voice of the theorist come through. I don't intentionally impose a scheme on the theories. If the theory is critical, I present it as critical, and if it is scientific, I present it as scientific.

There are two implications of this approach that I think are important. First, the book has many voices. It's *polyvocal,* which is part of what makes up the landscape of contemporary theory. The second implication is that I don't offer many critiques of these theories. Every theory in here is absolutely amazing! And the real thrill of theory is when that first lightbulb goes on. If I can get you to "get it" . . . well, let me just say that getting you to think a new thought is an ultimate high for me. It's what makes us most human, I think.

Obviously, the next step in theory is to be able to critique, but you should first have a lightbulb experience. They won't all hit you like this, but some of these theories will remake your world. In fact, it is my explicit intent to disturb and unsettle you. And the theories all by themselves will do that. However, the critiques of the theories are actually within the book. Each theory offers a different perspective than the one before it. For example, what is society? Is it a system, like Niklas Luhmann claims, or is it made up of population structures and exchange networks, like Peter Blau argues? Is the subject dead, as Jean Baudrillard claims, or is the self the central organizing feature of the encounter, as Erving Goffman claims? My point is that once you understand a theory, you will automatically think differently about other theories and form your own critiques.

The book also has some structural features I'd like to bring to your attention as well. Every chapter has an "Essential" box at the beginning of the chapter, and a "Building Your Theory Toolbox" feature at the end. The "Essential" box gives you a brief biographical sketch, as well as two other sections that I hope will give you a quick handle on what's going on: *Passionate Curiosity* (the central questions the theorist is interested in) and *Keys to Knowing* (central concepts to keep your eye open for).

The "Building Your Theory Toolbox" section has a number of resources to help you go beyond what's presented in the book. In *Learning More,* I list the more important primary and secondary works for an author. *Check It Out* highlights some of the more important or interesting topics from the chapter and points you to resources. A fairly consistent element of *Check It Out* is the *Web Byte* feature. Available on the book's Web site (http://www.pineforge.com/csstStudy), *Web Bytes* are relatively short yet substantial introductions to the work of 10 additional theorists. Using these *Web Bytes,* students and professors can further explore a theoretical issue raised in the main chapter; or, these *Bytes* can be used in concert with the book to emphasize certain areas of society or theory. For example, an entire course on inequality (class, status, and power) could be built by using Chapters 5, 6, 7, 8, 9, 13, 16, 17, and 18, coupled with the *Web Bytes* for *Stuart Hall and Reading Culture Through Cultural Studies; Candice West and Don H. Zimmerman: Doing Gender; Karen S. Cook: Power in Exchange Networks; Randall Collins and Conflict Theory; Erik Olin Wright: Measuring Class Inequality;* and *Patricia Hill Collins and Intersecting Oppressions.* This example is only one of many themes that can be developed. *Seeing the World* consists of review questions that you should be able to answer after studying the chapter. *Engaging the World* offers suggestions for using the theory; and, finally, *Weaving the Threads* are questions that ask you to compare, contrast, evaluate, critique, and synthesize certain ideas that two or more theorists have in common. These *Threads* should also sensitize you to the central issues in contemporary theory.

One more feature may be of interest to you. Important terms and concepts are either highlighted in *italics* or marked in **bold**. The terms in **bold** are defined in the online Glossary of Terms.

Acknowledgments

In any project of this type, there are many people that should be acknowledged for many different reasons. However, in the case of this book, one reason for appreciation and acknowledgment stands out from all the rest: patience. First, I want to acknowledge and thank my life-partner. Jen, you endured long days, missed weekends, and my dark hours of stress. And through it all you gave me support and laughter. I'll never be able to repay you, but I'll spend a lifetime trying. Next, I acknowledge and thank my editorial staff at Pine Forge: Benjamin Penner, Katja Werlich Fried, Annie Louden, and Laureen A. Shea. You all suffered patiently through missed deadlines, shifting tables of content, lost picture credits, monstrous word counts, and a list of inconveniences that would be longer than this book. You are the best; you are the A-Team! One editor deserves special mention when it comes to patience: my copy editor, Teresa Herlinger. Teresa, you've put up with grammatical and spelling errors that should have been eradicated in high school, with references that are either incomplete or superfluous, and with obtuse language that was better at hide-and-seek than at revealing. But your greatest gifts are your insights and questions; you keep me on my toes. Two other groups also gave patiently to this project: the faculty and students of the sociology department at the University of North Carolina, Greensboro. To the faculty—and especially the department head, Steve Kroll-Smith—my thanks for your patience over missed meetings and my infrequent presence in the department. To my students: Thank you for reading my work through its various iterations; thank you for listening to my rants and inevitable rabbit trails; and, most of all, thank you for taking a chance and having a thought. We are most human when we most clearly think; but, then, you already know that. I also want to thank the many, many reviewers who at one time or another contributed to this work. Without your input, this book would be substantially different. In particular, I thank Stephan Groschwitz (University of Cincinnati) for understanding and for critical and supportive input, and I thank Jeffery Ulmer (Pennsylvania State University) for specific recommendations for the organization of the book and ideas about symbolic interactionism. Finally, I want to thank Jerry Westby. You took the first chance and it's all your fault.

Pine Forge Press gratefully acknowledges the contributions of the following individuals:

John P. Bartkowski
Mississippi State University

Elena Bastida
The University of Texas Pan
    American

Bob Bolin
Arizona State University

Joseph Gerteis
University of Minnesota

Stephan F. Groschwitz
University of Cincinnati

Sarah Horsfall
Texas Wesleyan University

John A. Hughes
Lancaster University

Basil Kardaras
Capital University

Alem Kebede
California State University,
    Bakersfield

Eleanor A. LaPointe
Rutgers University, Georgian Court
    University, and Ocean County
    College

Shoon Lio
University of California–Riverside

James P. Marshall
University of Northern Colorado

Marietta Morrissey
University of Toledo

Glenn W. Muschert
Miami University

Yvonne Olivares
The Ohio State University

Frank J. Page
University of Utah

Dan Ryan
Mills College

Jeff Ulmer
Penn State University

Pablo Vila
Temple University

INTRODUCTION TO SECTION I

The Social Situation and Its People

I'd like you to think about the last time you got together with a group of your friends. What happened? What did you do? If you could answer me, chances are you would say something like "we just hung out and talked" or "we watched the football game on TV." Now, how would you answer that question theoretically? To put it another way, how would you answer the question in sociological terms? How does watching a football game relate to society? How can you think about hanging out and talking in theoretical terms? What is going on when we are face-to-face with other people? On the surface, these may seem like silly questions, but they are far from it.

In a broad way, the five theories presented in this section all look at what happens in and around social situations, like when you talk or watch TV with friends. The theories we'll be looking at don't present a complete understanding of what's going on in the situation; to present such a robust understanding would take a book of its own. In some cases, the theories reach beyond the situation, as with Berger and Luckmann's understanding of social reality and Randall Collins's theory of the micro–macro link. But together these theories give us a good place to begin thinking about the situation and its people.

The book begins with the idea of meaning. We use the term "meaning" quite a bit, but it's important for us to stop and take a look at its significance. First, everything human beings do is built around meaning. Our entire world is meaningful; things in our world may mean a lot or a little, but there's always meaning. Second, meaning is never the thing-in-itself. By definition, meaning is always something other than an event or object. And here's where things get interesting for us.

The idea of meaning gets played out in two ways in Section I. First, in Chapter 1, which deals with symbolic interaction, we'll see that meaning emerges out of interactions. If meaning is something other than the thing itself, then where is meaning and how can we know what it is? The answer symbolic interaction gives us is that meaning is negotiated through the back-and-forth interplay of verbal and nonverbal cues. That is, meaning emerges from the situation.

The second way that meaning is addressed in this section is through the social constructivist theory of Berger and Luckmann (Chapter 2). These two theorists argue that creating meaning is necessary for human survival. Humans are born instinctually underdeveloped within a world that has no specifically human environment. We use meaningful culture to both substitute for instincts and create and order a human world. There is, however, a problem: Meaning isn't real in the same way physical elements like mountains are real. So, how do we make something appear objectively real when it isn't? That's the topic of Berger and Luckmann's theory.

Chapter 1 also introduces us to the idea of the self. We will see that the self is an immanently social entity. That is, individuals don't need a self—society needs a self. The self is formed through successive stages of social role-taking in interaction. Through role-taking, the self is able to take the perspective of the generalized other. The self then works to provide control over individual behaviors, thus producing social order and making society possible.

In Chapter 4, Goffman takes the idea of the self in a different direction. Goffman isn't concerned about the "internalized self." Rather, he sees the self as the principal feature and guiding force in all social encounters. Situations demand that individuals present and manage a self for others to see and react to—it's the only way a social situation can occur. This implies two things for Goffman: First, the self is a dramatic effect that is more at home in the situation than the individual. And second, the situational requirement of presenting and managing a self imperceptivity produces the interaction order that undergirds all encounters. In other words, social order is an unintended consequence of impression management.

Like Goffman, Harold Garfinkel (Chapter 3) is concerned with social order. But unlike Goffman, Garfinkel doesn't consider the self at all; nor does he think about meaning or the interaction per se. What Garfinkel allows us to see is that we achieve social order in the most ordinary and unnoticed ways, as in saying "you know." Generally, if we ever think about such things as "you know," we think they are meaningless. But ask yourself this question: What does "you know" *accomplish* in an interaction? It doesn't *mean* something; it *does* something.

Randall Collins (Chapter 5) takes a different tack and claims that we are more concerned with diffuse emotional feelings than with meaning, order, or the self. According to Collins, it's much more likely that we feel our way to and through interactions, with our emotions working as a kind of radar and reservoir, rather than manage an impression or negotiate meaning. Further, Collins sees rituals as the most important part of the interaction. In rituals, we charge up our emotional energy and collect higher levels of cultural capital—at least, that's our intent.

Collins also provides a transition point into the next section of the book through the ideas of social structure and the micro–macro link. The issue of the micro–macro link is one that is obvious in contemporary theory and virtually unknown in classical theory. For some time, sociologists have thought about macro-level phenomena and micro-level interactions separately. In some ways, the two different domains seemed to discount one another. Micro-level theorists like George Herbert Mead saw social institutions more in terms of symbols and ways of thinking and behaving, with their importance and influence emerging out of interactions. On the other hand, structuralists such as Émile Durkheim saw human consciousness and behaviors as being the result of institutional arrangements. Eventually, sociologists began to see a theoretical issue here. If there are two separate fields, face-to-face interactions and social structures, then how are they related?

Collins provides us with a theory that links the micro level of the situation with macro-level processes. Simply, Collins argues that ritualized interactions are linked together through cultural capital and emotional energy. Individuals actually provide the link as they move from ritual to ritual, being drawn to those interactions where they are most likely to increase their holdings of cultural capital and emotional energy.

We'll see that Collins is able to explain such large-scale phenomena as stratification through his theory of interaction ritual chains. The interesting thing about Collins's theory, however, is that he argues against the existence of social structures. According to Collins (1975), "All social structure is problematic, quite possibly only a myth that people talk about or implicitly invoke when they encounter each other" (p. 53). For Collins, then, the situation is everything. What appear to be "structures" are in reality chains of interaction rituals.

So, what did you do the last time you got together with your friends? Maybe you didn't start a social movement, but in that mundane situation you negotiated symbolic meanings; engaged in self-talk and evaluation though role-taking; managed and maintained your self and the self of everyone else present; achieved a sense of social order and reality; participated in rituals that confirmed social reality; exchanged cultural capital and emotional energy; and linked up your one social situation with others occurring in the present, the past, and the future. Sounds like you were busy, and we've only made it through the introduction.

CHAPTER 1

Symbols, Meaning, and the Social Self

*George Herbert Mead (1863–1931)
and Herbert Blumer (1900–1987)*

Photo: Reprinted with permission of The Granger Collections.

Photo: Reprinted with permission of the American Sociological Association.

it is possible for a physical object to provide motivation, like when a rock falls on your hand. But something truly amazing happens not long after the rock hits you—you turn both it and the pain into social objects.

To really understand the issue behind symbols and social objects, let's consider natural signs. A sign is something that stands for something else, like smoke stands for fire. So, if we see smoke coming out of a room, we will call and report a fire, even though we may never actually see the fire. It appears that many animals can use signs. My dog, Gypsy, for example, gets very excited and begins to salivate at the sound of her treat box being opened or the tone of my voice when I ask, "Wanna treat?" But the ability of animals to use signs varies. For instance, a dog and a chicken will respond differently to the presence of a feed bowl on the other side of a fence. The chicken will simply pace back and forth in front of the fence in aggravation, but the dog will seek a break in the fence, go through the break, and run back to the bowl and eat. The chicken appears to only be able to respond directly to one stimulus, where the dog is able to hold his response to the food at bay while he seeks an alternative. This ability to hold responses at bay is important for higher-level thinking animals.

These signs that we've been talking about may be called *natural signs*. They are private and learned through the individual experience of each animal. So, if your dog also gets excited at the sound of the treat box, it is because of the dog's individual experience with it—Gypsy didn't tell your dog about the treat box. There also tends to be a natural relationship between the sign and its object, and these signs occur apart from the agency of the animal. In other words, Gypsy did not make the association between the sound of the box and her treats, I did. So, in the absolute sense, the relationship between the sound and the treat isn't a true natural sign. Natural signs come out of the natural experiences of the animal, and the meaning of these signs is determined by a structured relationship between the sign and its object.

Humans, on the other hand, have the ability to use what symbolic interactionists call significant gestures or symbols. In contrast to natural signs, symbols can be abstract and arbitrary. With natural signs, the relationship between the sign and its referent is natural (as with smoke and fire). But symbolic meaning can be quite abstract and completely arbitrary (in terms of naturally given relations). For example, the year 2006 doesn't exist in the physical world. What year it is depends on what calendar is used, and the different calendars are associated with political and religious issues, not nature (for examples, look up the Chinese and Muslim calendars on the Internet). Symbols are also reflexive, calling out the same response in the person speaking and listening. For example, I may call you up and say, "Tomorrow is Monday." Chances are that neither one of us will be happy about that situation. But while Monday is a social object and thus reflexive, it is also subject to interaction. Thus Monday becomes something totally different when I say, "And all classes are cancelled." Symbols, then, are verbal and nonverbal signals that convey meaning, require interpretation, and are reciprocal.

According to symbolic interactionists, the meaning that a symbol or social object cues is its set of organized sets of responses. In other words, symbolic meaning is not the image of a thing seen at a distance, nor does it exactly correspond to the dictionary, but the meaning of a word is the action that it calls out. For example, the meaning of a chair is the different kinds of things we can do with it. Picture a wooden object with four legs, a seat, and a slatted back. If I sit down on this object, then the meaning of it is "chair." On the other hand, if I take that same object and break it into small pieces and use it to start a fire, it's no longer a chair—it's firewood. So the meaning of an object is defined in terms of its uses, or legitimated lines of behavior. This definition of action is created through social interaction, both past and present.

Because the meaning, legitimated actions, and objective availability (they are objects because we can point them out as foci for interaction) of symbols are produced in social interactions, they are **social objects.** Any idea or thing can be a social object. Natural features such as the Smoky Mountains, invisible things like ghosts, and ideas like freedom can all be social objects. There is nothing about the thing itself that makes it a social object; an entity becomes a social object to us through our interactions around it. Through interaction we call attention to it, name it, and attach legitimate lines of behavior to it.

If you look around the room you are sitting in, everything you notice is a social object. In fact, there is a profound way in which people only see and relate to social objects. Human reality is constituted symbolically; it's a symbolic world, not a physical one, filled with social objects rather than physical objects. But can we notice things that aren't social objects? Yes, we can, but if we do, the object will be a problem for us to respond to because we won't know the meaning of the thing—note that the response itself becomes a social object (running from an unknown danger, investigating an unidentified entity, and so on).

Concepts and Theory: The Necessary Self

The self has the characteristic that it is an object to itself. . . . How can an individual get outside himself (experientially) in such a way as to become an object to himself? This is the essential psychological problem of selfhood or of self-consciousness. (Mead, 1934, pp. 136, 138)

The Self

Have you ever watched someone doing something? Of course you have. Maybe you watched a worker planting a tree on campus, or maybe you watched a band last Friday night. And while you watched, you understood people and their behaviors in terms of the identity they claimed and roles they played. In short, when you watch someone you understand her or him as a social object. After watching someone, have you ever called someone else's attention to that actor? Undoubtedly you have, and it's easy to do. All you have to say is something like, "Whoa, check her/him/it out." And the other person will look and usually understand immediately what it is you are pointing out, because we understand one another in terms of

Part of what we mean by the self is this internalized conversation. We carry on an internal dialog about who we are, what we are doing, where we are going, what the world means, and so on. This conversation is between the two parts of the self (the observer and the actor). Symbolic interactionism calls these two interactive facets of the self the "I" and the "Me." The Me is the self that results from the progressive stages of role-taking and is treated as a social object—it is that part of our self that observes our behaviors. The Me doesn't fully come into existence until we are able to role-take with the generalized other. The Me, then, is the perspective that we assume in order to view and analyze our own behaviors; the I is that part of the self that is unsocialized and spontaneous.

Remember what I mentioned a moment ago concerning what and where the self exists? Notice what we just said, the self is an internalized conversation between the I and the Me. And what do we already know about conversations and interactions? They are emergent. You and I may have a good idea of what we are going to talk about when we get together at the bar, but there's also a good chance that we'll end up talking about things we couldn't have imagined. Conversations shift and change, and meanings emerge through this negotiated interaction. And the same is true for the self. The self doesn't exist in any one place. It is a social object (something we give meaning to) that emerges through ongoing, internalized conversations and social interactions (role-taking). During the internalized interaction, it is specifically the I that makes the process emergent. The I is the actor, and it can act apart from the Me, apart from the perspective given by the generalized other.

Our *Web Byte* for this chapter, the work of R. S. Perinbanayagam, expands on this notion of the self being intrinsically tied to interaction—what Perinbanayagam calls "dialogic acts." Dialogic acts are those kinds of actions that are bound up with talking or dialog. Perinbanayagam's main point is that the self is specific to humans because human action is dialogic—it happens through and because of language and dialog. According to Perinbanayagam, it doesn't matter what we talk about, as long as we talk to each other. *It's conversation that gives presence to the self.* Specifically, people use rhetorical devices in order to be seen and noticed—apart from such devices, the self cannot exist. Perinbanayagam explores these and other ramifications of the dialogic self in our *Web Byte*. Check it out.

Social Action

So, why do you have a self? Symbolic interactionism teaches us that the self is created; it isn't something we're born with. In fact, there isn't anything within us as individuals that makes it natural or imperative that we have a self. Don't get me wrong, obviously there are internal preconditions that make the self possible and that predispose us toward having a self. But if it were possible for a human baby to grow to adulthood apart from society, that person wouldn't have, nor would she or he need, a self. In fact, in such a state the self would be a liability. Why? Because the self is not an intrinsic characteristic of the individual; it is a social entity. What we see from symbolic interactionism is that the self functions to control our behaviors—it allows us to act rather than react. A person who had lived alone for her or his entire existence would need to react instinctually to situations, rather than respond to their social meaning. Thinking about the social ramifications of some action would slow the beast down and make it vulnerable. Thus, the self is mandated socially, not individually, so that we can act rather than react.

Mead argues that *the act* contains four distinct elements: impulse, perception, manipulation, and consummation (behavior). We feel an impulse to behave: We are hungry, tired, or angry. For most animals, the route from impulse to behavior is rather direct—they react to the stimulus using instincts or behavioristically imprinted patterns. But for humans, it is a circuitous route. After we feel the initial impulse to act, we perceive our environment. This perception entails the recognition of the pertinent symbolic elements (including other people, absent reference groups, and so on) as well as alternatives to satisfying the impulse. *Perception* is the all-important pause before action; this is where society becomes possible. After we symbolically take in our environment, we manipulate the different elements. This manipulation takes place in the mind and considers the possible ramifications of using different behaviors to satisfy the impulse. We role-take with significant present and generalized others, and we think about the elements available to complete the task. After we manipulate the situation symbolically, we are in a position to act.

Notice how much of human action takes place in the mind. In fact, we could say that all the action that is distinctly human takes place in the mind. What I mean by that is that all animals have impulses and all animals behave. Those are the first and last stages of Mead's act. What makes behavior distinctly human are the middle two stages, both of which take place in the mind. And, guess what? The mind is social too. According to symbolic interactionists, the **mind** is a kind of behavior that involves at least five different abilities. It has the ability

1. To use symbols to denote objects

2. To use symbols as its own stimulus (it can talk to itself)

3. To read and interpret another's gestures and use them as further stimuli

4. To suspend response (not act out of impulse)

5. To imaginatively rehearse one's own behaviors before actually behaving

Let me give you an example that encompasses all these behaviors. A few years ago, our school paper ran a cartoon. In it was a picture of three people: a man and a woman arm-in-arm, and another man. The woman was introducing the men to one another. Both men were reaching out to shake one another's hands. But above the single man was a balloon of his thoughts. In it he was picturing himself violently punching the other man. He wanted to hit the man, but he shook his hand instead and said, "Glad to meet you."

There are a lot of things we can pull out of this cartoon, but the issue we want to focus on is the disparity between what the man felt and what the man did. He had an impulse to hit the other man, perhaps because he was jealous. But he didn't. Why didn't he? Actually, that isn't as good a question as, *how* didn't he? He was able to not hit the other man because of his mind. His mind was able to block his initial impulse, to understand the situation symbolically, to point out to his self the

Blumer isn't trying to do away with quantitative analysis either. His point is actually much more powerful. Quantitative data and statistical analysis can be used to great benefit in sociology, but sociologists need to be careful about what we think is actually influencing or producing an effect. Let's say we find a statistical association between gender and salary. For the symbolic interactionist, that association isn't the end of the study; it's just the beginning. If you were the researcher, you would then have to look for the empirical actions, interactions, and joint actions in back of the association between variables. In this sense, symbolic interactionism doesn't have a specific methodology. Almost all methodological approaches can be used if human agency and social interaction are given their proper place. (See Ulmer & Wilson, 2003, for a more complete introduction to SI and quantitative analysis.)

However, Blumer does give us two methodological recommendations: exploration and inspection. "Exploration is by definition a flexible procedure in which the scholar shifts from one to another line of inquiry, adopts new points of observation as his study progresses, moves in new direction previously unthought of, and changes his recognition of what are relevant data as he acquires more information and better understanding" (Blumer, 1969, p. 40). Exploration is grounded in the daily life of the real social group the investigator wants to study. Rarely does a researcher have first-hand knowledge of the social world she wants to study. Thus, rather than entering another's world with preconceptions, as much as is possible the researcher naively enters the other's world and searches for the social objects that the group regularly employs in producing their meanings through interactions. The records of such social objects and interactions become comprehensive and intimate accounts of what takes place in the real world. These accounts are in turn analyzed. The researcher seeks to sharpen the concepts she is using to describe the social world, to discover generic relationships (those that appear to hold true in various settings), and to form theoretical propositions.

Summary

- Human beings survive because of meaning, and meaning is something other than the thing or experience itself. Meaning is always symbolic and pragmatically oriented, and it emerges out of social interaction, a three-step interface of action: sending a symbolic cue, responding to the cue, and responding to the response. Generally speaking, social interactions do not stop at this point but continue on through many, many iterations of these three phases. Because interactions are ongoing, meaning is constantly emerging.
- The human world is made up of various social objects. Social objects come to exist as they are indicated by the interactants, and as particular kinds of actions are intended toward the thing. Social objects may be actual things, symbolic meanings, selves, or others.
- The self is a social object that is constructed through three stages of role-taking: the play, game, and generalized other. By symbolically seeing the self

from the role of the other, the person learns to divorce her- or himself from her or his own behaviors. A perspective is thus created—a place from which to view and attribute meaning to one's own behaviors. This perspective is referred to as the "Me," and the acting or impulsive side is called the "I." What we mean by the self only exists in the conversation between the I and the Me.

- The social object quality of the self allows people to consider and control their behavior. The self is what allows humans to act rather than react. A single act has four stages: impulse, perception, manipulation, and consumption. All animals have impulses—thus the distinctly human elements are found in the final three stages.

- Society thus does not determine our actions; action is a choice. There are two ways in which society exists and has influence: through institutions and as constructions of joint action. Institutions are potential sets of attitudes that interactants may role-take with; these sets of attitudes constitute the relevant generalized others. Joint actions are various individual interactions that are laced together.

- Blumer argues that the only empirical and acting part of society is the interaction, and he cautions us against the danger of reifying concepts such as institutions or social structures. In analyzing society, then, we need to focus on the interaction, realizing that it is an ongoing and moving process wherein individual actors exercise agency. This analysis should be in two phases. The first is exploration: Because of the emergent nature of society and self, researchers must divest themselves as much as possible from preconceived notions of what might be happening in any given situation. Theory ought to be grounded in the actual behaviors and negotiations in real interactions. Second, as theoretical concepts suggest themselves from the experience of the researcher in the field, these should be inspected to see if they might hold in other settings as well.

Building Your Theory Toolbox

Learning More—Primary Sources

- The foundational works for symbolic interaction are
  o George Herbert Mead: *Mind, self, and society: From the standpoint of a social behaviorist,* University of Chicago Press, 1934.
  o Herbert Blumer: *Symbolic interactionism: Perspective and method,* University of California Press, 1969.

Learning More—Secondary Sources

- For a good introduction to structural symbolic interactionism, look in the following:
  o Sheldon Stryker, *Symbolic interactionism: A social structural version,* Blackburn Press, 2003.

- Three important extensions of symbolic interactionism are
  - o *Affect control theory:* David Heise, Understanding social interaction with affect control theory, in *New directions in contemporary sociological theory* (J. Berger and M. Zelditch Jr., Eds.), Rowman & Littlefield, 2002.
  - o *Expectation states theory:* David. G. Wagner and Joseph Berger, Expectation states theory: An evolving research program, in *New directions in contemporary sociological theory* (J. Berger and M. Zelditch Jr., Eds.), Rowman & Littlefield, 2002.
  - o *Cultural studies:* Norman K. Denzin, *Symbolic interactionism and cultural studies: The politics of interpretation,* Blackwell, 1992.
- For a synthesized and general theory of interaction, see
  - o Turner, J. H., *A theory of social interaction,* Stanford University Press, 1988.

Check It Out

- *Web Byte—R. S. Perinbanayagam and Dialogic Acts*
- *The Society for the Study of Symbolic Interaction:* http://sun.soci.niu.edu/~sssi/
- *The Self:* For further reading, I would suggest that you read a standard approach—Morris Rosenberg, *Conceiving the self,* Basic Books, 1979—and a more postmodern reading—James A. Holstein and Jaber F. Gubrium, *The self we live by: Narrative identity in a postmodern world,* Oxford, 2000.

Seeing the World

- After reading and understanding this chapter, you should be able to answer the following questions (remember to answer them *theoretically*):
  - o What is pragmatism and what is its unique association with America? How does pragmatism inform symbolic interactionist theory?
  - o What is the importance of meaning and how is it achieved?
  - o How are social objects defined? What can be a social object? How are social objects used in interaction?
  - o What are the mind and self and why are they functionally necessary for society? What kind of behaviors do the mind and self engage in?
  - o How are the mind and self formed?
  - o What is the generalized other and what role does it play in self and society?
  - o What is society and how is it formed?

Engaging the World

- How would symbolic interactionists talk about and understand racial and gender inequality?
- Knowing what you know now about how the self is constructed, how do you think sociological counseling would look? Using your favorite Internet search engine, enter "clinical sociology." What is clinical sociology? What is the current state of clinical sociology?

- SI very clearly claims that our self is dependent upon the social groups with which we affiliate. Using SI theory, explain how the self of a person in a disenfranchised group might be different than one associated with a majority position.

Weaving the Threads

- At this point, all I can do is point out some ideas for you to be aware of as we work our way through this book. Begin thinking about the ideas of structure, symbols, language, meaning, self, and identity.

culture is taken as reality. Their focus, then, is on how culture presents itself to us in a way that appears objective and real. However, according to Stuart Hall and the school of cultural studies, there is something more going on with culture. Ask yourself, from where does most of our culture come? Does it come from the social groups to which we belong? Or is most of our culture created elsewhere, according to some other group's agenda? And, if our culture is created by others, and if, as Berger and Luckmann argue, culture is our reality, then whose reality are we living? How can we begin to look beneath the surface of our culture to see its political underpinnings? Cultural studies shows us a way through such ideas as hegemony, signs and semiotics, representation and discourse, and meaning and struggle. Take some time to check it out.

Objectivation: Making Meaning Real

With human beings, everything is possible but not all things are probable. The idea of the social construction of reality is a provocative one. It implies that reality is not simply there for us to discover; rather, it is made up, shaped, assembled, fashioned, formed, produced, and constructed. As such, the idea implies that everything is possible. If reality is simply a construction, then we can construct anything we like, right? As I said above, meaning and reality are changeable because they aren't tied to anything.

However, not all things are probable. The reason some things aren't probable is that reality is *socially* constructed. Reality isn't simply made up; it is achieved. In order for anything to become real for us, we must work collectively. Having others believe your reality is important, but there is more to reality construction. There is a method through which reality is attained. In other words, not only must we work collectively to achieve reality, we also have to do it in a certain way.

Berger and Luckmann call this reality work *objectivation.* Objectivation means to make something an object that isn't one. The feminist critique of patriarchy has for many years argued that men objectify women. In that case, we treat a human being that has feelings and values as if she were a sexual object, just a thing that exists for our pleasure, and in that context, it's offensive to many. However, objectivation of contingent meanings is necessary. Try and imagine a situation in which everybody knew and felt that everything they thought and believed was made up, fake, and imaginary. Such a scenario would, in Berger and Luckmann's (1966) apt description, lead to a nightmare of chaos: "*All* social reality is precarious. *All* societies are constructions in the face of chaos" (p. 103, emphasis original). Human action and interaction would be utterly impossible.

Meaning, because it is precarious and changeable, must be made to appear as if it were as stable, unquestionable, and taken for granted as any other thing in the environment. Once meaning has been objectified, it takes on the characteristics of facticity—meanings become facts. At one time in human history, it was a fact that the world was flat and that the earth was the center of the universe. From a constructivist point of view, the facticity of the flat earth and the facticity of the round globe are achieved in the exact same ways. Whether or not one of them is "true" doesn't matter and in some ways isn't discernable. Humans in every case make the meanings they construct appear real and objective. Ideas become facts not because they are, but because we make them appear as such through objectivation.

There are four major ways in which we objectify culture: institutionalization, historicity, legitimation, and through language. **Institutionalization** begins as behaviors become habitual. In every situation, human beings need to figure out how to act. We don't have instincts that direct our actions; our actions are meaningful and thus directed by culture. When we enter a new situation, we need to determine what kind of situation it is and what kind of behaviors are meaningfully relevant—it would be silly to act like a student while on a date; student behaviors such as rising your hand and note taking do not have the same meanings on a date as they do in the classroom. If we had to figure out how to act in every situation from scratch, we would soon be paralyzed. It would take us far too much time to process all the necessary data and contingencies that each situation would bring. So, humans habitualize their actions—we make much of our behavior a matter of routine.

Institutionalization of activities occurs "whenever there is a reciprocal typification of habitualized actions by types of actors" (Berger & Luckmann, 1966, p. 34). If something is reciprocal, it means that it is shared or things correspond to one another. Humans also make actions and people typical in every circumstance. That is, rather than seeing an action as unique and a person as utterly individual, we see them as types. For example, when I walk down the hall at my school and see a student sitting on the floor looking through a pile of 3x5 cards, I think, "Ah, she is studying for a test." And even if I know that woman is Stephanie, I still see her as a type of person: a student studying for a test. The behavior is typical for that type of person. Further, when Stephanie looks up at the person passing her, she sees me, a professor on his way to class. Thus, the habitual actions and person types are reciprocally held by both me and Stephanie, and by everybody else across the United States in that type of situation. They are what "everybody knows" about school or any other situation. In other words, these reciprocal typifications and routines are there for us all. They present themselves to us as objective even though they are constructed.

Berger and Luckmann argue that humans will always tend to reciprocally habitualize and typify. Even if we start with two people in a new situation, over time they will begin to expect typical behaviors from one another. To illustrate this point, Berger and Luckmann give us a thought experiment. Let's use their illustration and pretend that you and I have never met and that we hold no typifications. This is an impossible situation because as soon as you see me you'll type me as a male; but for the sake of this thought experiment let's pretend that there are no typifications. Let's say that we are going to build a boat. We've never built a boat but we know we need to. Each day we come together and work on this boat. At first, all our actions and reactions are tentative, because we don't know what to expect from one another or from this kind of work. But eventually things smooth out and we are able to work much faster. What makes things smooth out is that I figure out what kinds of things you do in the process of building the boat and you reciprocally figure out what kinds of things I do. We come to know what to expect because we know what types of actions the other is likely to perform. We type each other vis-à-vis our boat behaviors.

In this imaginary world, let's pretend that you're green and I'm blue. The things that I do around the boat, then, are green behaviors and yours are blue. We are able to predict one another's behaviors because our actions and interactions have become habitual and typical; this enables us to save time and energy. The work becomes more efficient, and we are then able to come up with innovations to our boat-building process, which subsequently become part of our routines and division of labor.

Now we need to add one more element: children. Let's say we have two blue children and two green children. What kind of boat behaviors will we expect from the kids? Chances are we won't start out with a blank slate. We will expect green behaviors from the green children and blue behaviors from the blue children. We will, in fact, go one step further. We'll teach the different types of children the behaviors that are typical for that type of person. This is much easier and more efficient than trying to come up with new meanings for each new person.

Notice what happens with this new generation. You and I were there from the beginning. We have subjective memories of our boat-building experiences. The world of boat building is completely transparent to us—we can see it for all it is. For the children, however, the boat world is a very different place. It isn't transparent to them; it existed before they were born, along with its typifications. They have no subjective memory of boat building; it is completely objective. *Historicity* develops, and as more generations come and go, "the objectivity of the institutional world 'thickens' and 'hardens,' not only for the children, but (by mirror effect) for the parents as well. . . . A world so regarded attains a firmness in consciousness; it becomes real in an ever more massive way and it can no longer be changed so readily" (Berger & Luckmann, 1966, p. 59). Thus, generally speaking, the longer the history attached to reciprocal relations, the greater the objectivity of the institution.

Something else happens here as well. As we've seen, humans demand meaning, and the children will want to know what all children want to know: why? Why are we building this boat? What does the boat mean? Why am I expected to do green work? Why can't I play with Bobbie, just because he is blue? The children and grandchildren and great-grandchildren don't have a memory of boat building, yet they need to know its meaning. Rather than a biographical memory of the world, subsequent generations get *stories* about the boat world. These stories are legitimations. **Legitimations** are stories that give social and power relationships a cognitive and moral basis. Social worlds need legitimation because humans need meaning. And as the history and objectivation of an institution increase, so do the legitimating stories. They become more robust and more complex as time and generations go by.

There are different levels of legitimation. Some legitimations are self-evident: Just by the nature of its existence, the behavior or social relation is legitimated. Berger (1967) gives us a crude but effective example: "'You ought not to sleep with X, your sister'" (p. 30). Why can't I sleep with that woman? Because she is your sister. We perceive that there is something inherent in that relationship that needs no further explanation: It's obvious; how could you even think it? *Self-evident legitimations* are present in any statement of "that's the way things are."

Let's take this crude example through the different levels of legitimation. If I continue to demand an explanation for why I can't sleep with my sister, you might offer a more *theoretical legitimation*. You might say, "Sex between siblings is one of the basic taboos of existence. We are genetically predisposed to react with revulsion to sibling sex because the offspring of such a union would weaken the gene pool." This legitimation moves beyond the self-evident and offers a cognitive, well-reasoned argument about the prohibitions of the relationship.

There is yet a more powerful level of legitimation. This is the *legitimation of symbolic universes*. The theoretical explanation that was given above is quite specific. It is a genetic theory concerning probable chromosomal abnormalities that might occur as the result of sibling sex. The theory says nothing about gay marriage, or mass media, or the thousands of other things that make up our lives. However, symbolic universes do just that. Symbolic universes are systems of meaning that embrace *all existence* into a single meaningful whole. These symbolic systems offer explanations and legitimations for all the things and experiences that affect human beings. The clearest example of such a system of meaning is religion. "Religion implies that human order is projected into the totality of being. Put differently, religion is the audacious attempt to conceive of the entire universe as being humanly significant" (Berger, 1967, p. 28).

One of the functions of legitimations is to make the socially constructed world appear natural and not the result of human agency. And no knowledge system does this as well as religion in all its forms. There are four specific "gains" in reality construction from using religion as legitimation. First, human institutions are seen as manifestations of the underlying structure of the universe. Berger and Luckmann refer to this as an equation between the *nomos* and *cosmos*—the nomos being human order and the cosmos being the order of the universe. Religion is able to lift humanly produced institutions totally out of their contingency and bestow "upon them an ultimately valid ontological status, that is, by locating them within a sacred and cosmic frame of references" (Berger, 1967, p. 33).

Let's take the question of gay marriage as an example. If marriage is a human institution that was constructed during the Middle Ages in part to assure the patriarchic control of property, then the case for not allowing gay marriage is rather weak. The historical model makes marriage a human institution created to achieve specific social goals. Thus, not allowing gays to marry is merely society deciding that it doesn't want to extend certain civil rights to particular groups of people. However, if marriage is an institution created by God to be a reflection of the relationship between Christ and his bride, the church, then marriage between a man and a woman is part of the eternal scheme of the universe and humankind cannot change it. This justification function appears in all religions throughout history and across societies, from the justification of the Crusades to the justification of the caste system (social position as the result of reincarnation). Religion makes the contingency of the human situation seem ultimately real.

The second gain from religious legitimation is that it defines disorder as evil. As we've seen, humans create order. We impose order on geography by creating state and national boundaries; we impose order on the endless variety of human

CHAPTER 3

Organizing
Ordinary Life

Harold Garfinkel (1917–)

Photo: Reprinted with permission of Bernard Leach, Manchester Metropolitan University.

However, indexical expressions are reflexive for another reason. They not only appear in and reference the situation; they also bring the situation into existence. Mehan and Wood (1975) give us the example of "hello." Let's say you see me in the hall at school. You say, "Hello." What have you done? You have initiated or created a social situation through the use of a greeting. When you said "hello," you immediately drew a circle around the two of us, identifying us as a social group distinct from the other people around us. That social situation, which we can call an encounter, interaction, or situated activity system, didn't exist until you said "hello." But notice something very important about "hello": It can only exist as a social greeting within social situations. Every time "hello" is used, a social situation is created. Yet "hello" is only found in social situations, either real ones or imaginary ones (like with our example). Thus, "hello" is utterly reflexive: It simultaneously creates, exists, and finds meaning within the social situation.

It isn't just "hello" that exists reflexively. Let's take the phase, "you're beautiful." Its meaning is obviously contextual. You might say "you're beautiful" to a queen, to your partner after making love, to your friend who just made a particularly ironic comment, or to your friend dressed up to go to a Halloween party. But notice also that saying "you're beautiful" also creates the situation wherein beautiful is understood. The beauty of the ironic comment didn't exist until you said it; once said, it can be understood within the context that it created.

Further, indexical expressions aren't limited to these sorts of catch phrases. At one point, Garfinkel asked his students to go home and record a conversation. They were to also report on the complete meaning of what was said. The following is a small snippet of one such report (Garfinkel, 1967, pp. 25–26):

	What was said:	*What was meant:*
Husband:	Dana succeeded in putting a penny in a parking meter today without being picked up.	This afternoon as I was bringing Dana, our four-year-old son, home from the nursery school, he succeeded in reaching high enough to put a penny in a parking meter when we parked in a metered parking zone, whereas before he has always had to be picked up to reach that high.
Wife:	Did you take him to the record store?	Since he put a penny in a meter, that means that you stopped while he was with you. I know that you stopped at the record store either on the way to get him or on the way back. Was it on the way back, so that he was with you, or did you stop there on the way to get him and somewhere else on the way back?

The first thing to point out, of course, is that what was actually said is incomprehensible apart from what the members could assume the other knew. There is an entire world of experience that the husband and wife share in the first statement about Dana that gives the statement a meaning that any observer would not be able to access. So, the first point is apparent: Vocal utterances reference or index presumed shared worlds.

The second point may not be quite so obvious. The students had a difficult time filling out the far right column. It was hard to put down in print what was actually being said and indexically understood. However, it became a whole lot tougher when Garfinkel asked them to indexically explain what was said in the far right column. Garfinkel wanted them to explain the explanation because the explanation itself assumed indexical worlds of meaning. Garfinkel (1967) reports that "they gave up with the complaint that the task was impossible" (p. 26). The task of explaining every explanation is impossible because all our talk is indexical. Many of us come up against this issue in the course of raising a 2-year-old. All 2-year-olds are infamous for asking the same insistent question: "Why?" And every parent knows that once started, that line of questioning never ends—every answer is just another reason to ask why. It never ends because our culture is indexical and reflexive.

Mehan and Wood (1975) further point out just how fundamentally reflexive our world really is. Every social world is founded upon incorrigible assumptions and secondary elaborations of belief. **Incorrigible assumptions** are things that we believe to be true but never question. These assumptions are incorrigible because they are incapable of being changed or amended. And these assumptions form the base of our social world. **Secondary elaborations** of belief are prescribed legitimating accounts that function to protect the incorrigible assumptions. In other words, secondary elaborations of belief are ready-made stories that we use to explain why some empirical finding doesn't line up with our incorrigible assumptions. The really interesting thing about incorrigible assumptions is that the empirical world doesn't always line up with the cultural assumptions that guide and create our reality.

Mehan and Wood (1975) give us the illustration of a lost pen. Our search for the lost pen is based on the assumption of object consistency—physical objects maintain their consistency through time and space. "Say, for example, you find your missing pen in a place you know you searched before. Although the evidence indicates that the pen was first absent and then present, that conclusion is not reached" (p. 12). To do so would challenge the incorrigible assumption upon which that reality system is based—we never consider the possibility that a poltergeist took the pen or that a black hole swallowed it up. Our assumption of object consistency is protected through secondary elaborations of belief. When we find the pen where we had already looked, we say, "I must have missed it."

In Figure 3.1, I've pictured two models of sociology. These models obviously are not theoretical or causal but are simply pictures of assumptions and activities. The first model is what most sociologists see themselves as doing. Most sociologists assume that there is an empirical, social world that sociological methods can be used to study. These methods produce a particular kind of inquiry known as sociology. This kind of inquiry leads to insights into and discoveries about the social refuse to permit each other to understand what we are really talking about. It isn't that we are being intentionally deceitful; it is simply that behind every social activity lie endless fields of indexical worlds. So we use ad hoc measures to preserve a sense of shared social worlds. We let things pass and wait for clarification of statements that never comes. And we say things like "you know" (known as the *et cetera principle*) to indicate that we could explain if need be, but "you know" (even though what we all "know" never appears).

Radical Reflexivity

I'd like to offer one last word about Garfinkel's ethnomethodology and the idea of the reflexivity of social worlds. You might have noticed that ethnomethodology seems to challenge almost every authoritative or legitimated perspective or way of seeing the world. As we talked about earlier, all knowledge systems are based on incorrigible assumptions that are protected by secondary elaborations of belief. And as we saw in the story of my first teaching assistantship, sociology itself is understood by ethnomethodologists as a reflexively produced way of seeing the social world. This method is termed **endogenous reflexivity:** "Endogenous reflexivity refers to how what members do in, to, and about social reality constitutes social reality" (Pollner, 1991, p. 372). This is what we were talking about above in Figure 3.1: What sociologists do to make their work appear sociological creates the very reality they are looking at.

Pollner (1991) points out that there are two other levels of reflexivity in ethnomethodological analysis, referential and radical reflexivity. In *referential reflexivity*, the researcher turns her gaze to her own methods: Every analysis and knowledge system is reflexively produced, including ethnomethodology. Part of the ethnomethodological stance, then, is to see one's own way of understanding the social world as being reflexively produced. Referential reflexivity is radicalized when the researcher not only understands her way of knowing as reflexively produced but also sees her own self or identity within that system as being reflexively produced. If, for example, I'm an ethnomethodologist, in *radical reflexivity* I would see both my knowledge and my identity as an ethnomethodologist as being reflexively produced.

Why would ethnomethodology want to push this idea of reflexivity so far? What can be gained when everything is cut loose from any empirical reality? Just this: "Intrinsic to radical reflexivity is an 'unsettling,' i.e., an insecurity regarding the basic assumptions, discourse and practices used in describing reality" (Pollner, 1991, p. 370). I wonder what we could see and learn if everything were unsettled. One thing is for certain: In making everything unsettled, ethnomethodologists want us to see how it is done in just this way at just this time.

One final word about ethnomethodology and my experiences with Eric Livingston. About midway through the semester, I felt I was beginning to figure it out, and I was intrigued. One day while we were walking together to lunch, I asked him if we could get together for a couple of sessions. I wanted him to teach me ethnomethodology. And Eric very much wanted me to learn ethnomethodology. But, he told me, it wasn't something that could be taught through reading or talking. We would need to spend many hours simply watching people—watching them at the malls, crossing streets, lining up at the movies, attending classes, and doing the thousand-and-one things that people do each and every day.

I was a little surprised. I thought I had asked a regular question, kind of like asking Herbert Blumer to explain symbolic interactionism to me. And I was disappointed. I was carrying a heavy academic load in school and was a single parent; I just didn't have the time to invest in that kind of learning process. Plus, Eric was only going to be at the university for 1 year, so it wasn't something that we could schedule for some other time. It all added up to our just having a few more conversations on the subject, but no people watching.

It didn't really strike me then, but I can see now why Eric said we needed to spend much more time learning this theoretical perspective: Ethnomethodology is more caught than taught. Probably more than any other perspective in this book, ethnomethodology must be applied to be known. As I mentioned before, I think the chief difficulty with this perspective is that it deals with those things that are seen but unnoticed. We miss the power of ethnomethods because they are so commonplace—and they are powerful precisely because they are ordinary.

Ethnomethodology really is more caught than taught. So quit reading and go watch people. And ask yourself one simple question: What is this behavior or part of a conversation *doing?* Don't ask what it *means;* ask, rather, *what does this do?* Begin to think about interactional elements as mechanisms that achieve something in the social encounter. If you continually ask yourself this question, you'll soon begin to become aware of the seen but unnoticed foundations of social order.

Summary

- Garfinkel's perspective is unique among sociologists. He sees social order and meaning as achievements that are produced *in situ.* That is, Garfinkel sees social order and meaning as achieved *within its natural setting*—face-to-face interactions—and not through such things as institutions that exist outside the natural setting.

- The principal way this is done is through accounting. A basic requirement of every social setting is that it be recognizable or accountable as a specific kind of setting. Thus, the practical behaviors that create a setting just as it is are seen but not noticed for what they are; they are the very behaviors that achieve the setting in the first place.

- All settings and talk are therefore indexical; they index or reference themselves. The actions that create the situation of a wedding or a class are simultaneously understood as meaningful, social activities within the situation. Human activity always references itself; it is thoroughly reflexive, based upon incorrigible assumptions, discovered through the documentary method, proven through indexical methods, and protected by secondary elaborations of belief.

CHAPTER 4

Performing the Self

Erving Goffman (1922–1982)

Photo: Courtesy of the University of Pennsylvania Department of Sociology.

What's going on?

Dateline: Friday, May 2, 2003. **ABOARD USS ABRAHAM LINCOLN (CNN)**—President Bush made a landing aboard the USS Abraham Lincoln Thursday, arriving in the co-pilot's seat of a Navy S-3B Viking after making two fly-bys of the carrier.

It was the first time a sitting president has arrived on the deck of an aircraft carrier by plane. The jet made what is known as a "tailhook" landing, with the plane, traveling about 150 mph, hooking onto the last of four steel wires across the flight deck and coming to a complete stop in less than 400 feet.

Moments after the landing, the president, wearing a green flight suit and holding a white helmet, got off the plane, saluted those on the flight deck and shook hands with them. Above him, the tower was adorned with a big sign that read, "Mission Accomplished." (CNN, 2003)

What's going on?

One day while driving to a lunch appointment, I stopped at a stop light. Like in most cities at that hour, the majority of cars were filled with business people going to lunch. In front of me was a man in a black BMW with his windows rolled down. While I watched, he changed his shirt, coat, and hat, and he put on cologne (I could smell it after he sprayed). He also switched radio stations, from a news station to a hip-hop station.

W hat's going on in these scenarios may seem clear. In each case, the person is working at presenting a specific image that is meant to convey a certain kind of self. I chose the two examples in order to highlight the fact that this kind of impression management occurs in both formal and informal settings. However, Erving Goffman says there's actually more going on than might first meet the eye.

In Chapter 1, about symbolic interaction (SI), we saw that the self is important for controlling one's own behaviors, which is vital for society. But the self as a social object, as perceived by others, is secondary; it may or may not become part of the interaction. If it becomes a focus, its meaning and substance are negotiated and emergent. SI is also quite concerned about the internalization process: it's a social psychological perspective that sees internalization occurring as the individual role-takes. Successive role-taking experiences produce the "Me" component of the self.

Goffman sees the self differently from this. First, Goffman isn't concerned about the communication between the "I" and the "Me." He changes the focus from internal reflexivity to the demands made by the encounter. Second, according to Goffman, the self isn't one of many possible social objects; *the self is the central organizing feature of all social encounters.* Further, we'll see that this presentation of self is more complex than we might think. We, like the president or the man in the BMW, may want to give off a certain impression, but how we do that and the ramifications of doing it are vast and generally unseen.

Third, Goffman doesn't focus on the internalization process. He probably wouldn't argue with symbolic interaction about this issue, but Goffman's analytic focus is the interaction order. Everything about the self, then, is related to this way of seeing the encounter. Thus, all the selves and identities that we have are put together out of the dramatic realization of the interaction. Goffman argues that these things become internalized not because of role-taking, but because of situational factors directly related to the presentation of self.

The Essential Goffman

Biography

Goffman was born on June 11, 1922, in Alberta, Canada. He earned his Ph.D. from the University of Chicago. For his dissertation, he studied daily life on one of the Scottish islands (Unst). The dissertation from this study became his first book, *The Presentation of Self in Everyday Life,* which is now available in 10 different languages. In 1958, Herbert Blumer invited Goffman to teach at the University of California, Berkeley. He stayed there for 10 years, moving to the University of Pennsylvania in 1968, where he taught for the remainder of his career. Goffman also served as president of the American Sociological Association in 1981 and 1982.

Passionate Curiosity

Goffman was inquisitive about everything people did in face-to-face interactions. He watched them continually. The face-to-face encounter so enthralled Goffman that he was driven to probe the interaction itself: What are the requirements of an encounter? How do these requirements influence everything that people do when they meet?

> **Keys to Knowing**
>
> interaction order, impression management, front, teams, setting, front and back stages, role distance, deference and demeanor, face-work, focused and unfocused encounters, frames

Goffman's Perspective: All the World Is a Stage

Dramaturgy: Performing the Self

Goffman's perspective has become known as dramaturgy. **Dramaturgy** is a way of understanding social encounters using the analogy of the dramatic stage. In this perspective, people are seen as performers who are vitally concerned with the presentation of their character (the self) to an audience. There are three major premises to dramaturgy. First, all we can know about a person's self is what the person shows us. The self isn't something that we can literally take out and show people. The self is perceived indirectly through the cues we offer others. Because of this limitation, people are constantly and actively involved in the second premise of dramaturgy: impression management. *Impression management* refers to the manipulation of cues in order to organize and control the impression we give to others. We use staging, fronts, props, and so on to communicate to others our "self" in the situation, and they do the same for us.

Taken alone, impression management sounds at best strategic and at worst deceitful. If everybody around us is manipulating cues in order to present a specific self, then how can we believe that it is their "true self" that we see? The short answer is that we can never be sure that we see an authentic self; we always have to *assume* that the self we see is real. But notice something here: *For the interaction, it does not matter if the self we see is genuine or false*—whether authentic or fake, all selves are communicated in the exact same way, through signs that are specifically given or inadvertently given off.

The third premise of dramaturgy is that there are particular features of face-to-face encounters that tend to bring order to interactions. The presentation of self places moral imperatives on interactions. Selves are delicate things and are easily discredited. If you have ever felt embarrassed, you know the painful reality of this truth. Selves depend upon not only our skill in presenting and maintaining cues, but also the willingness and support of others. Thus, the simple act of presenting a self creates a cooperative order.

The Interaction Order

In many ways, Goffman is not so much interested in the interaction per se, as he is in *the order that is demanded by interaction*. The content of the interaction, its meanings and motivations, isn't Goffman's concern. For us to interact, there are certain rules and ways of behaving that are demanded. These ways of behaving have little if anything to do with our personal motives, but have significant power over the effects of the interaction. For example, when we enter an interaction, we find out what people are talking about, what kinds of roles are important, what statuses are claimed, how involved people are in the interaction, and so on. Then, once we've checked out the terrain, we gradually introduce our talk and self into the flow of interaction. The motive behind such care, Goffman tells us, is to save our self from embarrassment. The effect, however, is that the organization of the interaction is preserved: "His aim is to save face; his effect is to save the situation" (Goffman, 1967, p. 39).

Goffman, however, is quick to tell us that he is not proposing a situational reductionism, where the only thing that exists is face-to-face interactions. Goffman doesn't talk in terms of structures or institutions, but he names at least three things that exist outside of the immediate interaction. First, *settings* strongly inform the definition of the situation. The definition of the situation is important because it tells us what kind of selves to present, what to expect from others, how to interpret meaning, and so forth. For example, the selves, meanings, and others available in a university classroom are different than at a local bar.

Another element that exists outside the encounter is *biographies*. People come into interactions with biographies. These biographies are previously established stories about our self and others. There are two kinds of biographical stories that we use, individual and categoric. If we see someone with whom we have interacted previously, then we have an individual biography of that person and she or he of us. As we will see, these biographies or story lines are the result of impression management. Yet, once established, you and I are both committed to maintaining that story. When a personal biography isn't available, as when you first meet someone, then categoric biographies are used—stories that go along with the type of person you are meeting. For example, when you first meet a professor, there's a categoric biography that you access, a story about that person even though you've never met before. Both individual and categoric biographies structure the encounter.

The third extra-interactional element that Goffman (1983) explicitly talks about is *cognitive relations:* "At the very center of interaction life is the cognitive relation we have with those present before us" (p. 4). As members of categoric groups, each of us has identifiable knowledge bases, and these islands of knowledge are related to other specific categoric groups. For example, let's say it's Friday after work and you've just stopped by a local bar to unwind. As you sit down at the bar, the person next to you strikes up a conversation. The small talk continues as you chat about work, the poker tournament on the television, and other bits and pieces of social life. Then one of you mentions music and you find out that one plays guitar and the other drums. Suddenly an entire horizon of shared knowledge opens up. You can almost feel the expansion from a narrow sliver of shared reality to a world of cognitive relations.

Thus, Goffman isn't arguing that more macro-level entities don't exist, or that they are any less an abstraction than the interaction order (though you might notice that the way he talks about such large-scale things is distinctly different from most structural sociologists). It's Goffman's (1983) perspective that "in all these

There have been a few really great minds. Among artists we can think of Dali, da Vinci, Matisse, Monet, Picasso, and so on. Einstein, Darwin, Galileo, Curie, and Newton number among the scientists. Philosophy's pantheon includes such people as Plato, Descartes, Kant, Wittgenstein, and Heidegger. In our own discipline we have Marx, Weber, and Durkheim. These are people so prominent that most of us only need their last names to recognize them. They are towering intellects one and all. But have you ever thought that there might be a sociological component to their thinking and creativity? Randall Collins thinks there is.

Shifting gears, let's think for a moment about 9/11. There is little doubt that the events of that day have influenced life in America. But what was it that changed America? The easy answer is that the attacks themselves impacted us. However, think simply about the physical attributes of the terrorist acts. The explosions leveled buildings, yet there was a material boundary where physical destruction stopped. Sound waves continued past that point and people heard the explosions but did not experience any damage. Past the point where sound stopped, there were little if any physical effects. Nonetheless, most people in the United States "felt" the attacks, and we continue to feel their reverberations. What is it that we sense? How is it that an event that occurred hundreds, even thousands of miles away can impact us so strongly? Yet have you noticed that while 9/11 still affects us, it doesn't have nearly the power it did for the first few months after the attack? How can we understand the initial impact and then its waning? Is there a single theory that can explain both intellectual creativity and the impact of 9/11? Again, Randall Collins thinks there is.

Collins gives us an abstract theory that can explain such diverse issues and more. And it is a theory that provides a clear link between macro-level phenomena (like the U.S. response to 9/11) and micro-level interactions (like intellectual creativity). For a theory to propose that these two apparently dissimilar levels of social phenomena are related, it must have a specific mechanism that links one level to the next. Collins provides such a link—as well as an explanation of intellectual creativity and the nationalist response to 9/11. For Collins, the link is provided through chains of interaction rituals, specifically through the emotional energy and cultural capital that are created within encounters and then transferred to the next interaction. One note before we go further: The breadth of Collins's theory is larger than we can cover in this chapter. We won't, for example, be covering his conflict or geo-political theories. You will find those aspects of Collins's theory in the form of *Web Bytes*.

The Essential Collins

Biography

Randall Collins was born in Knoxville, Tennessee, on July 29, 1941. His father was part of military intelligence during WWII and then a member of the state department. Collins thus spent a good deal of his early years in Europe. As a teenager, Collins was sent to a New England prep school, afterward studying at Harvard and the University of California, Berkeley, where he encountered the work of Herbert Blumer and Erving Goffman, both professors at Berkeley at the time. Collins completed his Ph.D. in 1969. He has spent time teaching at a number of universities, such as the University of Virginia, the Universities of California at Riverside and San Diego, and has held a number of visiting professorships at Chicago, Harvard, Cambridge, and at various universities in Europe, Japan, and China. He is currently at the University of Pennsylvania.

Passionate Curiosity

Collins has enormous breadth, but seems focused on understanding how conflict and stratification work through face-to-face ritualized interactions. Specifically, his passion is to understand how societies are produced, held together, and destroyed through emotionally rather than rationally motivated behaviors.

Keys to Knowing

interaction rituals, interaction ritual chains, emotional energy, cultural capital, deference and demeanor rituals, order-takers and order-givers, bureaucratic personality, ritual density

Collins's Perspective: Science, Emotion, and Exchange

Scientific Theory

Collins (1986) is a social scientist who sees theory at the core: "The essence of science is precisely theory . . . a generalized and coherent body of ideas, which explain the range of variations in the empirical world in terms of general principles. . . . more centrally, it is explicitly cumulative and integrating" (p. 1345). As such, Collins's theorizing is cast at fairly high levels of generalizability. This is particularly noteworthy when he talks about emotion and culture. When Collins talks about emotion, he never talks about specific emotions like love, joy, hate, and so forth. It is emotion *generally* with which Collins is concerned. The same is true of his interest in culture. In Collins's hands, culture becomes symbolic goods that are used in exchange or sacred symbols that unite a group. Again, it is culture in the broadest sense of the word that Collins has in mind.

The kind of theory cumulation that Collins engages in is creative synthesis. In a synthesis, we take elements from different theories and bring them together to form a new, hopefully more powerful theory. One of the most notable things about the way Collins synthesizes theory is his thriftiness. As we work through his theory, you'll see that Collins takes a basic principle from Durkheim (the production of the sacred), modifies it with an insight from Goffman (interaction ritual), and blends it with Weber's idea of legitimacy. From this simple recipe, Collins creates a theory with very few elements but with enormous explanatory power. As Collins (1975) notes about theory, "a good theory gives a coherent vision within which research can elaborate complexities without having them overwhelm us" (p. 51).

Exchange and Emotion

Collins (1993) is an interesting kind of exchange theorist: He sees emotion as the common denominator of rational action. To bring emotion in, he points to three long-standing critiques of exchange theory. First, exchange theory has a difficult time accounting for altruistic behavior. Merriam-Webster (2002) defines altruism as "uncalculated consideration of, regard for, or devotion to others' interests." If most or all of our interactions are exchange-based and if all our exchanges are based on self-motivated actors making rational calculations for profit, how can altruism be possible? Collins claims that exchange theorists are left arguing that the actor is actually selfish in altruistic behavior—she or he gains some profit from being altruistic. However, just what that profit is has generally been left unspecified.

Second, there is evidence that suggests that people in interactions are rarely rational or calculative. In support of this, Collins cites Goffman's and Garfinkel's work, the idea of bounded rationality in organizational analysis, as well as psychological experiments that indicate that when people are faced with problems that should prompt them to be rational, they use non-optimizing heuristics instead. These heuristics function like approximate or sufficient answers to problems rather than the most rational or best answer. The third criticism of exchange theory is that there is no common metric or medium of exchange. Money, of course, is the metric and medium of trade for exchanges involving economically produced goods and services; however, money isn't general enough to embrace all exchanges, all goods, and all services.

Collins sees each of these problems solved through the idea that *emotional energy* is the common denominator of rational action. Let's note from the beginning that this approach is rather adventuresome in that it combines two things that have usually been thought of as oil and water—emotion and rationality just don't mix. At least, they didn't before Collins came along. **Emotional energy** does not refer to any specific emotion; it is, rather, a very general feeling of emotion and motivation that an individual senses. It is the "amount of emotional power that flows through one's actions" (Collins, 1988, p. 362). Collins (2004a) conceptualizes emotional energy as running on a continuum from high levels of confidence, enthusiasm, and good self-feelings to the low end of depression, lack of ambition, and negative self-feelings (p. 108). The idea of emotional energy is like that of psychological drive, but emotional energy is based in social activity.

Collins is arguing that emotional energy is general enough to embrace all exchanges. In fact, emotional energy is the underlying resource in back of every exchanged good and service, whether it's a guitar, a pet, a conversation, a car, a friend, your attendance at a show or sporting event, and so on. *More basic than money, emotional energy is the motivation behind all exchanges.* Emotional energy can also be seen in back of social exchanges that might seem counterintuitive. Why would I exchange my free time to work at a soup kitchen on Sunday mornings? This, of course, is an example of altruistic behavior. Exchange theory, apart from the idea of emotional energy, is hard pressed to explain such behaviors in terms of exchange. Collins gives us a more general property of exchange in the form of emotional energy. People engage in altruistic behaviors because of the emotional energy they receive in exchange.

The idea of emotional energy also solves the problem of the lack of rational calculations. As Collins notes, people aren't generally observed making rational calculations during interactions. Rather than being rationally calculative, "human behavior may be characterized as emotional tropism" (Collins, 1993, p. 223). A *tropism* is an involuntary movement by an organism that is a negative or positive response to a stimulus. An example of tropism is the response of a plant to sunlight. The stems and leaves react positively to the sun by reaching toward it, and the roots react negatively my moving away from it and deeper in the ground. Collins is telling us that people aren't cognitively calculative in normal encounters. Instead, people emotionally feel their way to and through most interactions, much like a plant reaches toward the sun.

The *Web Byte* for this chapter—the work of Thomas J. Scheff—takes the idea of emotion in a different direction: the effects of pride and shame on the individual. Like Collins, Scheff gleans inspiration from Goffman; but unlike Collins, Scheff brings his insights inside the person. Scheff's theory is also like Collins in that it is based on emotion, but Scheff has particular kinds of emotion in mind, where Collins speaks of emotion in a general sense. Where Collins uses deference and demeanor rituals to theorize about stratification, Scheff sees deference as the core issue in all interactions. One further comparison: Both Scheff and Collins give us theories to explain creative genius; but, again, Collins tunes us into those factors that exist in the situation and Scheff focuses on the person. In some interesting ways, Collins and Scheff represent two sides of the same coin. Collins provides us with the situational effects and Scheff the inner effects of social emotion.

The Situation

In simple terms, there are four possible sites of social and behavioral research: the individual, the interaction or social situation, social structures, and social or global systems. Collins makes a strong case for starting with the social situation rather than the individual, structures, or systems. He points out that what we mean by the individual varies by the social and cultural context and is thus a poor focus for social science research. There are two ways to understand his point. The first is to understand that what we mean by "the individual" is really the social point at which various social identities meet. For example, if I were to ask you to tell me who you are, most of your answers would be in the form of social categories and would involve such things as age, gender, sexuality, friendship, marital status, and so on. The individual, from this point of view, is a reflection of sociopolitical organization rather than essential characteristics.

The other way to see that the individual varies by social context is much more profound. From this perspective, the entire idea of "the individual" is the product of political, religious, and social changes that have occurred in the past few centuries. More specifically, the idea of the individual came about as Western society defined civil rights (as a result of the rise of democracy) and moral responsibilities (as a result of the Protestant Reformation). The idea of the individual also became more pronounced through capitalism (consumerism) and social diversity. About the individual, Collins (2004a) says, "The human individual is a quasi-enduring, quasi-transient flux in time and space. . . . It is an ideology of how we regard it proper to think about ourselves and others . . . not the most useful analytical starting point for microsociology" (p. 4).

On the other side of the situation are structures. As we've seen, the idea of social structures is held in question by most of the theorists in the first section of this book (in Section II, we will be introduced to a set of theorists for whom structure is central in theorizing about society). For Collins, social structures and systems are *heuristics*—that is, they are aids to discovery. Collins (1987) is arguing that we can use the ideas of structure and social systems to "make generalizations about the workings of the world system, formal organizations, or the class structure by making the appropriate comparisons and analyses of its own data" (pp. 194–195). But the reality behind these heuristics is the pure number of face-to-face situations strung out over time and space. In other words, *social structures are built up by the aggregation of many interactions over long periods of time and large portions of geographic space.* Here we find part of Collins's micro–macro link, which we will soon explore.

Concepts and Theory: Interaction Ritual Chains (IRCs)

Rituals

For Collins, **rituals** are patterned sequences of behavior that bring four elements together: bodily co-presence, barrier to outsiders, mutual focus of attention, and shared emotional mood. These elements are variables—as they increase, so also will the effects of ritualized behavior. There are five main effects of interaction rituals: group solidarity, group symbols, feelings of morality, individual emotional energy, and individual cultural capital. Collins's theory is diagrammed in Figure 5.1.

One of the first things that the model in Figure 5.1 calls our attention to is physical *co-presence*, which describes the degree of physical closeness. Even in the same room, we can be closer or further away from one another. The closer we get, the more we can sense the other person. As Durkheim (1912/1995) says, "The very act of congregating is an exceptionally powerful stimulant. Once the individuals are gathered together, a sort of electricity is generated from their closeness and quickly launches them to an extraordinary height of exaltation" (pp. 217–218). Bodily presence appears theoretically necessary because the closer people are, the more easily they can monitor one another's behaviors.

Part of what we monitor is the level of involvement or *shared focus of attention,* the degree to which participants are attending to the same behavior, event, object, symbol, or idea at the same time (a difficult task, as any teacher knows). We watch bodily cues and eye movements, and we monitor how emotions are expressed and how easily others are drawn away from an interaction. Members of similar groups have the ability to pace an interaction in terms of conversation, gestures, and cues in a like manner. Part of the success or intensity of an interaction is a function of this kind of rhythm or timing.

The key to successful rituals "is that human nervous systems become mutually attuned" (Collins, 2004a, p. 64). Collins means that in intense interactions or ritual

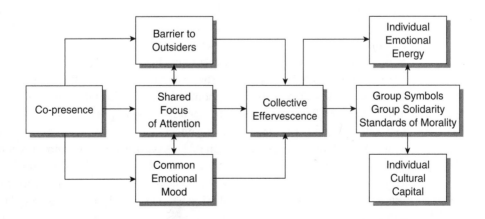

Figure 5.1   Basic Interaction Ritual

performance, we physically mimic one another's body rhythms; we become physically "entrained." *Rhythmic entrainment* refers to recurrent bodily patterns that become enmeshed during successful rituals. These bodily patterns may be large and noticeable as in hand or arm expressions, or they may be so quick and minute that they occur below the level of human consciousness.

It is estimated that human beings can perceive things down to about 0.2 second in duration. Much of this entrainment occurs below that threshold, or below the level of consciousness—which indicates that *people literally feel their way through intense ritualized interactions.* Collins (2004a, pp. 65–78) cites evidence from conversational analysis and audience–speaker behavior to show that humans become rhythmically coordinated with one another in interactions. Some research has shown that conversations not only become rhythmic in terms of turn-taking, but the acoustical voice frequencies become entrained as well. EEG (electroencephalogram) recordings have indicated that even the brain waves of interactants can become synchronized. In a study of body motion and speech using 16mm film, Condon and Ogston (1971) discovered that in interaction, "a hearer's body was found to 'dance' in precise harmony with the speaker" (p. 158).

A shared focus of attention and common emotional mood tend to reinforce one another though rhythmic entrainment. *Common emotional mood* refers to the degree to which participants are emotionally oriented toward the interaction in the same way. In ritual terms, it doesn't really matter what kind of emotion we're talking about. What is important is that the emotion be commonly held. Having said that, I want to point out that there is an upper and lower limit to ritual intensity. One of the things that tends to become entrained in an interaction is turn-taking. The rule for turn-taking is simple: One person speaks at a time. The speed at which turns are taken is vitally important for a ritual. The time between statements in a successful conversation will hover around 0.1 second. If the time between turns is too great, at say 1.0 second, the interaction will be experienced as dull and lifeless and no solidarity will be produced. If, on the other hand, conversational statements go in the other direction and overlap or interrupt one another, then the "conversation" breaks down and no feeling of solidarity results. These latter kinds of conversations are typically arguments, which can be brought about by a hostile common emotional mood. Collins (2004a) points out that it is generally the case at the micro level that "solidarity processes are easier to enact than conflict processes. . . . The implication is that conflict is much easier to organize at a distance" (p. 74).

Being physically co-present tends to bring about the other variables, particularly, as we've seen, the shared focus of attention. Co-presence also aids in the production of ritual barriers. *Barrier to outsiders* refers to symbolic or physical obstacles we put up to other people attempting to join our interaction. The use of barriers increases the sense of belonging to the interaction that the participants experience. The more apparent and certain that boundary, the greater will be the level of ritualized interaction and production of group emotional energy. Sporting events and rock concerts are good illustrations of using physical boundaries to help create intense ritual performance.

Notice that there is a total of five effects coming out of the emotional effervescence that is produced in rituals. Let's first talk about the interrelated group effects first: solidarity, group symbols, and standards of morality. *Group symbols* are those symbols we use to anchor social emotions. The greater the level of collective effervescence that's created in rituals, the greater will be the level of emotion that the symbol comes to represent. It's this investment of group emotion that makes the symbol a collective symbol. The symbol comes to embody and represent the group. If the invested emotion is high enough, these symbols take on sacred qualities. We can think of the United States flag, gang insignia, and sport team emblems and colors as examples, in addition to the obvious religious ones. The symbols help to create group boundaries and identities. Group symbols have an important ritual function: They are used to facilitate ritual enactment by focusing attention and creating a common emotional mood.

Group solidarity is the sense of oneness a collective can experience. This concern originated with Durkheim (1893/1984, pp. 11–29) and meant the level of integration in a society, measured by the subjective sense of "we-ness" individuals have, the constraint of individual behaviors for the group good, and the organization of social units. Collins appears to mean it in a more general way. Group solidarity is the feeling of membership with the group that an individual experiences. It's seeing oneself as part of a larger whole. One of the important things to see here is that the sense of membership is emotional. It is derived from creating high levels of collective effervescence. Of course, the higher the level of effervescence, the higher will be the sense of belonging to the group that an individual can have.

Standards of morality refer to group-specific behaviors that are important to group membership and are morally enforced. Feelings of group solidarity lead people to want to control the behaviors that denote or create that solidarity. That is, many of the behaviors, speech patterns, styles of dress, and so on that are associated with the group become issues of right and wrong. Groups with high moral boundaries have stringent entrance and exit rules (they are difficult to get in and out of). Today's street gangs and the Nazi party of WWII are good examples of groups with high moral boundaries.

One of the things to notice about our example is the use of "moral." Most of us probably don't agree with the ethics of street gangs. In fact, we probably think their ethics are morally wrong and reprehensible. But when sociologists use the term moral, we are not referring to something that we think of as being good. A group is moral if its behaviors, beliefs, feelings, speech, styles, and so forth are controlled by strong group norms and are viewed by the members in terms of right and wrong. Because the level of standards of morality any group may have is a function of their level of interaction rituals, we could safely say that, by this definition, both gangs and WWII Nazis are probably more "moral" than we are, in this sense, unless one of us is a member of a radical fringe group.

The Micro–Macro Link

In Collins's theory of interaction ritual chains, the individual is the carrier of the micro–macro link. There are two components to this linkage: emotional energy and cultural capital. Emotional energy is the emotional charge that people can take away with them from an interaction. And as such, emotional energy predicts the likelihood of repeated interactions: If the individual comes away from an interaction with as high or higher emotional energy than she or he went in with, then the person will be more likely to seek out further rituals of the same kind. Emotional energy also sets the person's initial involvement within the interaction. People entering an interaction that are charged up with emotional energy will tend to be fully involved and more readily able to experience rhythmic entrainment and collective effervescence.

Cultural capital is a shorthand way of talking about the different resources we have to culturally engage with other people. The idea of cultural capital covers a full range of cultural items: It references the way we talk; what we have to talk about; how we dress, walk, and act—in short, anything that culturally references us to others. Collins lists three different kinds of cultural capitals. *Generalized cultural capital* is the individual's stock of symbols that are associated with group identity. As Figure 5.1 notes, a great deal of this generalized cultural capital comes from interaction rituals. This kind of cultural capital is group specific and can be used with strangers, somewhat the way money can. For example, the other day I was in the airport standing next to a man wearing a handmade tie-dye tee-shirt with a dancing bear on it. Another fellow who was coming off a different flight saw him and said, "Hey, man, where ya from?" These two strangers were able to strike up a conversation because the one man recognized the group symbols of Deadheads—fans of the band, The Grateful Dead. They were able to engage one another in an interaction ritual because of this generalized cultural capital.

Particularized cultural capital refers to cultural items we have in common with specific people. For example, my wife and I share a number of words, terms, songs, and so forth that are specifically meaningful to us. Hearing Louis Armstrong, for instance, instantly orients us toward one another, references shared experiences and meanings, and sets us up for an interaction ritual. But if I hear an Armstrong song around my friend Steve, it will have no social effect—there are no shared experiences (past ritual performances) that will prompt us to connect. From these two examples, you get a good sense of what cultural capital does: It orients people toward one another, gives them a shared focus or attention, and creates a common focus of attention, which are most of the ingredients of an interaction ritual.

The last kind of cultural capital that Collins talks about is *reputational capital.* If somebody knows something about you, she or he is more likely to engage you in conversation than if you are a complete stranger. That makes sense, of course, but remember that this is a variable. Mel Gibson, for example, has a great deal of cultural capital. If he were seen in a public space, many people would feel almost compelled to engage him in an interaction ritual.

Figure 5.2 depicts Collins's idea behind interaction ritual chains. Notice that there are several interactions pictured and that each interaction is made up in this case of two people. Each person comes into an interaction with stocks of emotional energy (EE) and cultural capital (CC) that have been gleaned from previous interactions. The likelihood of an individual seeking out an interaction ritual is based on his levels of emotional energy and cultural capital; the likelihood of two people interacting with one another is based on both the similarity of their stocks and

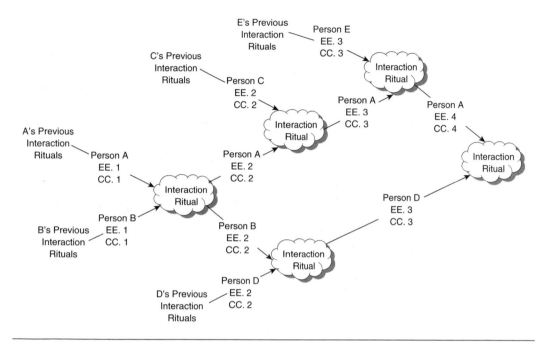

Figure 5.2   Interaction Ritual Chains

the perceived probability that they might gain either emotional energy or cultural capital from the encounter.

The micro–macro link for Collins, then, is created as individual carriers who are charged up with emotional energy and cultural capital seek out other interaction rituals in which to revitalize or increase their stocks. If you follow Person A through Figure 5.2, what you see is a trail of interactions that may form the beginning of an interaction ritual chain. As this chain builds up over time and space, a macro structure is formed.

One further point before we leave this idea: People have a good idea of their *market opportunities,* which is directly linked to cultural capital and indirectly to emotional energy. As we noted earlier, we exchange cultural capital in the hopes of receiving more cultural capital back. Part of our opportunity in the cultural capital market is structured: Our daily rounds keep us within our class, status, and power groups. However, our interpersonal markets are far more open in modern societies than they were in traditional ones. In these open markets, we are "rational" in the sense that we avoid those interactions where we will spend more cultural capital than we gain, and we will pursue those interactions where we have a good chance of increasing our level of cultural capital. We also tend to avoid those interactions where our lack of CC will be apparent. As a result, we tend to separate ourselves into symbolic or status groups.

and between social units, like people, organizations, and other structures. Pay close attention and you'll see that the theorists in this section emphasize different elements of this definition.

In Chapter 6, Peter Blau will introduce us to exchange theory. The first sociologist to work with the concept of exchange was Georg Simmel (1895–1918), but exchange didn't appear again in the sociological landscape until the publication of George Homans's *Social Behavior: Its Elementary Forms* in the early 1960s. Homans's approach was noteworthy in that he borrowed heavily from psychological behaviorism, a rather unique approach in sociology. Blau published not long after Homans and in my opinion represents a more sociological approach to exchange.

There are two additional reasons I've chosen to work with Blau over some of the other exchange theorists. First, Blau gives us a clear theory of a micro–macro link. You recall from our discussion of Randall Collins that this issue of the connection between situational and structural processes is an important area in contemporary theory. However, where Collins sees interaction ritual chains as the dynamic of the micro–macro link, Blau sees the link arising out of exchange processes that result in social structures (not chains of interactions). The second reason to use Blau is that in the latter part of his career, he abandoned exchange theory and its understanding of social structures for a different approach to structure: population structures. Rather than being built up through social exchange and norms, as with social structures, population structures are based on the physical distribution of populations in space. Thus, with Blau we actually get two different views of structure.

In Chapter 7, we will consider structures of inequality, specifically gender. The argument here, of course, is that social structures enable and constrain differently based on group membership—to state the obvious, social structures enable whites and males and constrain people of color and females. Chafetz's approach is interesting in that she argues that the economic structure plays the most significant role, but she also includes structuring forces at the micro and meso levels.

There are three additional reasons for me to use Chafetz at this point. Two of the reasons have to do with the way Chafetz theorizes. She draws from a variety of theories to synthesize a new, more powerful understanding. In doing so she (1) allows us to consider a number of gender theorists rather than just one, and (2) her work is a very good example of the sociological approach to theory building: cumulation, building on the work of others. The third reason for me to include Chafetz is that she not only explains how gender is produced and structured, she also explains how it can change. This is one of the more powerful features of sociological theory. If we know how something works, we are in a much better position to see how it can be fixed (if we think it's broken).

Another significant structure of inequality is class. Pierre Bourdieu is our theorist here. However, you will find Bourdieu's approach to be different from most other treatments of class. First, he gives us a brand new way of understanding and theorizing structure. Bourdieu is the first in this book to seriously challenge the dichotomies of agency–structure and micro–macro links (another such theory is Giddens's structuration theory in Chapter 12). Bourdieu terms his method of theorizing "constructivist structuralism," an approach that sees structure and agency producing a creative tension. The second distinctive feature of Bourdieu's theory is where he locates the structures that replicate class—principally in the human body.

With Bourdieu we also come to our first neo-Marxist theory. The preface "neo" means new; thus, neo-Marxist theory generally refers to updated versions or extensions of Marx's theory. In this general sense, Wallerstein's theory (Chapter 9) is neo-Marxist. But neo-Marxism also has a more specific meaning: It distinctly refers to a group of theoretical works that are based more on Marx's theories of consciousness and culture than on his work on class, the economy, and the state. Having said that, it's important to note that neither classic nor neo-Marxists reject Marx's material or cultural work; it's simply a matter of emphasis. Bourdieu is a good example here: He is concerned with material class but spends much more time explaining the impact of cultural issues, such as the internalization of class-based culture and symbolic violence. As we'll see in Chapter 11, Jürgen Habermas's approach is another example of the more specific understanding of neo-Marxism.

I also want to point out that we have our first neo-functionalist in this section: Luhmann. According to Jeffrey C. Alexander (1985), neofunctionalism "indicates nothing so precise as a set of concepts, a method, a model, or an ideology. It indicates, rather, a tradition" (p. 9). What he means is that neofunctionalism is more of a perspective than a precise theory. It's a way of looking at or analyzing the social world that has several identifying characteristics.

Most importantly, neofunctionalism gives us a way of seeing the interrelations of social parts. It is a way of describing the parts of the social system as symbiotically linked to one another and interacting without direction from an outside force. Neofunctionalism, then, encourages us to not look at social phenomena as if they exist in a vacuum. Rather, sociological inquiry ought to always consider the wider context and the relationships that might exist between our research topic and the other elements in society. It's this idea of connections between structures that brings in the dynamics of systems.

Actually, both Wallerstein and Luhmann work from a systems point of view, although they are concerned with different kinds of systems. Systems are defined in terms of relatively self-contained wholes that are made up of variously interdependent parts. That's a mouthful, but it really isn't as daunting as it might seem. Your body is a system, and it's made up of various parts and subsystems that are mutually dependent upon one another (your lungs would have a hard time getting along without your stomach, for example), each part contributing to the working of the whole. In society, the parts that we are talking about are usually conceived of as social structures.

Luhmann approaches the study of society from a neofunctionalist perspective, but even in that he is somewhat unique. He doesn't define systems by the interrelationships among structural parts, as is the usual approach; rather, Luhmann sees systems in terms of the boundary between a system and its environment. We can again use the analogy of the body to see this issue. Human bodies exist separately from but dependent upon the environment. There's a definite place where your body ends and the world around you begins, yet your body takes from and gives back to this environment. It's a simple idea, but, as you'll see, it has powerful consequences in Luhmann's hands—society, interactions, and individuals become utterly different from everything we've seen before.

Wallerstein's approach should feel a bit more familiar. He uses specific elements of Karl Marx's theory to argue that the capitalist economic system is now worldwide rather than isolated within the bounds of any one nation-state and territory. As you may recall from a course on classical theory, Marx argued that there are certain dialectical processes within capitalism, such as overproduction, that will eventually lead to the downfall of capitalism. Thus, one of the critiques of Marx's theory has been that the revolution hasn't occurred and capitalism is alive and well. In seeing capitalism as a world system, Wallerstein counters this critique: Capitalism hasn't yet failed because it has been able to overrun its national boundaries and displace its contradictions outside the state-level economy. Now that capitalism is the world system, Wallerstein argues, many of the dynamics that Marx theorized are taking hold.

CHAPTER 6

Social and Population Structures

Peter M. Blau (1918–2002)

Photo: Courtesy of Judith Blau.

I magine, if you would, a world where everything you want is yours. It doesn't matter what it is; you have inalienable rights to everything in the world, and every need and desire are fulfilled because there are no restrictions on you. In this world you are alone, but loneliness isn't a problem; you're not even aware that other people are possible. (Keep this in mind as we go through out little thought experiment: In terms of emotional needs and ties, relationships aren't an issue.)

Now, imagine that same world but with two differences: There is one other person, and the "rights to everything in the world" are divided equally. At this point in your world, things shift dramatically. The most profound change is that there is the possibility of relationship. But, again, this relationship isn't emotional; it's rational. It is centered on goods and resources. In this case, the possibility of a relationship happens in two ways.

First, you may want something that the other person controls. If so, you have three options: do without it, steal it, or work out an exchange. With all three options, you will engage in a calculation of costs and benefits. There are costs associated with doing without, with stealing it, and with exchanging some good or service for what you want. The issue in each case is this: Do the benefits outweigh the costs? But there's another issue that may not be as apparent: Only one of the options maintains individuality, and that's doing without. The other two options establish some kind of relationship with the other. And here is where costs versus benefits can get a little tricky.

If you decide to steal it, you may get caught and it then becomes a conflict. On the other hand, if you decide to exchange, then there are two possible complications: equity in exchange and delayed gratification. Granted, these complications only become important if there is the possibility of other exchanges in the future; but it seems reasonable to assume that if you need one thing from this other person, you'll need more. So, what happens if the exchange isn't equal? This could happen for any number of reasons, but for the moment let's just assume that there is inequity. What are the ramifications of such a situation?

The issue of delayed gratification concerns timeliness (or asynchronous exchanges). Let's suppose that you own the rights to all the cattle and the other person owns the rights to all the trees. In this scenario, she or he wants to eat and you want to build a house. The problem is that the trees aren't mature; they aren't big enough to cut for lumber, but the cattle are ready now. And the other person needs to eat now. In this situation, you either don't exchange or you give the person the cattle now with the promise of lumber later. What fundamental social quality is required for this situation to occur successfully? The answer, as you can probably guess, is trust.

The second way a relationship could occur is if there are needs that can only be met through cooperation. Let's keep this simple and say that you both need a large boulder moved. You decide to work together to move it. What problem might arise? In order to see the potential problem more clearly, let's say that the world is divided up among six rather than two people, and you all decide to move the boulder. Now you have six people moving the boulder. What problem could exist?

It's an interesting thought experiment, and if you play around with it for awhile you might be able to tease out some more interesting implications about human beings. But the most important thing that we need to see right now is that *these very basic exchange relationships have unintended consequences.* In each of our scenarios, the aim is simply to acquire a good or meet a need, yet each of the possible relationships we explored implies additional features. The ramification of inequity in exchange is social power; the fundamental feature necessary for asynchronous exchange is trust; and the possible issue arising from the need being met through cooperation is the problem of free riders (people who reap the benefits without incurring the costs—one of the six could simply *pretend* to lift the boulder).

Peter Blau does one of the best jobs of clearly explaining how these unintended consequences come together to form society. As such, he gives us our first taste in this book of how structures are built and our second answer to the problem of the micro–macro link; Randall Collins gave us our first answer. Blau uses elements that come directly from exchange theory to explain how structures are built: in particular, general trust and the norm of reciprocity. Blau also gives us a theory of power based on exchange principles (for other theories of power, see the chapters on Collins, Bourdieu, Foucault, Smith, and Butler). However, Blau's idea of structure is unique: He gives us a theory of *population structures* rather than social structures.

Social structures in sociology are usually thought of in terms of status positions, roles, norms, and other types of social factors. Population structures, on the other hand, are concerned with the actual demographics or statistical properties of the population, such as the relative size of groups, the number of different experience influence choice and preference), and the actual properties of exchange. Blau argues that exchange is an emergent property of interaction that cannot be reduced to the psychological attributes of individuals.

The idea of exchange as a social dynamic began with *utilitarianism,* an 18th-century philosophy that came out of the Age of Enlightenment's concern with human happiness and scientific calculation. The principles of utilitarianism were most clearly stated by Jeremy Bentham (1789/1996, pp. 11–16), who argued that happiness and unhappiness are based on the two sovereign masters of nature: pleasure and pain. The "utility" in utilitarianism refers to those things that are useful for bringing pleasure and thus happiness. Bentham developed the *felicific* or *utility calculus,* a way of calculating the amount of happiness that any specific action is likely to bring. The calculus had seven variables: intensity, duration, certainty/uncertainty, propinquity/remoteness, fecundity, purity, and extent. Knowing or memorizing Bentham's calculus isn't important; what is important is that Bentham introduced the idea of rational calculation being used to decide human behavior and the moral status of any act.

For Blau, exchange is an elementary process of human life and the prototype of social phenomena. By definition, exchange can only take place in social settings between two or more people. Social exchange focuses on the actions of the participants and how they are influenced by both the anticipated and past actions of others. The influence, then, is decidedly social and not based on personal achievements, such as education, or individual attitudes. However, Blau claims, not all face-to-face interactions are exchanges. This is a rather unique position among exchange theorists, many of whom see exchange or rational choice as universal to all human action. Blau (1968), on the other hand, sees that "the concept of exchange loses its distinctive meaning and becomes tautological if all behavior in interpersonal relations is subsumed under it" (p. 453). Other factors he sees as influencing people in interactions are morals in the form of internalized norms, irrationality (purely emotional responses), and coercion.

Social exchanges are distinct from economic exchanges in at least four ways. First, they lack specificity. All economic exchanges take place under the contract model. In other words, almost all of the elements of the exchange are laid out and understood in advance, even the simple exchanges that occur at the grocery store. Social exchanges, on the other hand, cannot be stipulated in advance; to do so would be a breach of etiquette. Imagine receiving an invitation for dinner that also stipulated exactly how you would repay the person for such a dinner ("I'll give you one dinner for two lunches"). Social exchanges, then, cannot be bargained and repayment must be left to the discretion of the indebted.

This first difference implies the second: Social exchanges necessarily build trust, while economic exchanges do not. Since social exchanges suffer from lack of specificity, we must of necessity trust the other to reciprocate. This implies that relationships that include social exchange—and almost all do—build up slowly over time. We begin with small exchanges, like calling people on the phone, and see if they will reciprocate. If they do, then we perceive them as worthy of trust for exchanges that require longer periods of time for reciprocation, such as friendship.

The third difference between social and economic exchanges is that social exchanges are meaningful. The way we are using it here, meaning implies signification. In other words, an object or action has meaning if it signifies something beyond itself. What we are saying about exchange is that a purely economic exchange doesn't mean anything beyond itself; it is simply what it is: the exchange of money for some good or service. Social exchanges, on the other hand, always have meaning. For an example, let's take what might appear as a simple economic exchange: prostitution. If a married man gives a woman who is not his wife money for sex, it is a social exchange because it has meaning beyond itself: In this case, the meaning is adultery.

Finally, the fourth difference between social and economic exchanges is that social benefits are less detached from the source. We use money in economic exchanges, but the value of money is completely detached from the person using it. I may use money every time I go to the music store, but the value of that money for exchange is a function of the United States government and has nothing to do with me—the value of money is completely detached from me. However, the value in all social exchanges is dependent upon the participants in some way. For example, the social exchange between you and your professor requires you to fulfill the requirements of the course to get a grade. Someone else can't do the work and you can't legitimately buy the grade.

Taken together, these four unique features of social exchange create diffuse social obligations. For example, let's say you and your partner invite another couple over for dinner. You expect that the invitation will be reciprocated in some way, but exactly how the other couple is to reciprocate isn't clear—nor *can* it be clear; to make it clear would reduce it to an economic exchange. So you have a general, unspecified expectation that the other couple will reciprocate in some way. The reciprocation has to be in the indefinite future (the other couple can't initially respond to your invitation by scheduling their "repayment" dinner—then it would really look like a repayment in economic terms), yet it has to be repaid specifically by the couple (the other couple can't have a different couple invite you for dinner and have it count for them). The dinner is meaningful, but the meaning isn't clear as of yet (Are you all going to be friends? If so, what kind of friends?). Thus, you have to trust the other couple to provide the future unspecified meaning and reciprocation. The other couple is obligated to you, but in a very diffuse manner.

Population Structure and Social Contact

As I mentioned previously, Blau argues that there are two broad influences on human behavior: those that influence choices and preferences and those that restrict or enable the actions that come from the choices and preferences. In contrast to many sociologists, Blau generally talks about population structures rather than social structures. With the idea of **population structures,** Blau is specifically interested in the distribution of a population among social positions and the extent to which the positions are correlated. We can think of these social positions as distinctions or differences that the population makes within itself. Race, for example, is a social position on a single dimension within the population structure.

have high value. However, repeated profits of the same kind have declining value. For example, the first time a man gives flowers to his partner, it has high value. And it probably has high value the second and third times it happens. But if the man brings flowers to his partner every Friday after work, the value of the gift declines as the partner becomes satiated. Thus, the value of any good or service is higher if there is some degree of uncertainty or a sporadic quality associated with it.

Exchange Norms and Social Power

There are two norms associated with social exchange: the norm of reciprocity and the norm of fair exchange. Blau sees exchange as the starting mechanism for social interaction and group structure. Before group identities and boundaries, and before status positions, roles, and norms are created, interaction is initiated in the hopes of gaining something from exchange. That we are dependent upon others for reaction implies that the idea of *reciprocation* is central in exchange. The word *reciprocate* comes from Latin and literally means to move back and forth. Exchange, then, always entails the give and take of some elements of value, such as money, emotion, favors, and so forth. As such, one of the first behaviors to receive normative power is reciprocity. The norm of reciprocity implies another basic feature of society, that of trust. As we've seen, social exchange requires that the reciprocated good or service be unspecified and that reciprocation is delayed to some undisclosed future. This lack of specificity obviously demands trust, which forms the basis of society and our initial social contact. Exchanges are also guided by *the norm of fair exchange*. Something is fair, of course, if it is characterized by honesty and free from fraud or favoritism. The expectation of fairness increases over the length of the exchange relation.

Because of the lack of specificity in social exchange and the norm of reciprocity, exchange creates bonds of friendship and establishes power relations. The basic difference between friendship and power relationships concerns the equity of exchange and is expressed in the amount of repayment discretion. Friendship is based on an equal exchange relationship: All parties feel that they give about as much as they take in the relationship. This equal reciprocity among friends leads to a social bond built on trust and a high level of discretion in repayment. On the other hand, inequality in exchange leads to unfulfilled obligations that, in turn, grant power over repayment to the other.

According to Blau, there are four conditions that affect the level of social **power.** I've diagramed these conditions in Figure 6.1. In the diagram, we have a social exchange relation between A and B. In thinking through the way the model works, you can visualize yourself as either person and get a sense of the way the power flows in the relationship. Social capital refers to the ability to participate in an exchange with goods or services that the other desires. As you can see, there is a negative relationship between social capital and power. In other words, the less ability Actor A has to control goods and services (exchange capital) that Actor B desires, the greater will be B's power over A. Alternatives are important for power as well. If I have a large number of alternatives through which I can obtain the social good or

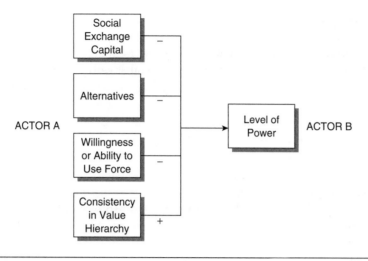

Figure 6.1   Exchange Principles of Power

service that I desire, then others will have little power over me. However, the fewer the number of alternatives, the greater will be the power of others who control the social good. This inverse relationship is noted by the negative sign.

There are two other important factors in establishing social power: the actor's willingness to use force and the consistency of the value hierarchy. If Actor A has the ability and willingness to use force, then by definition there is no exchange relation and, thus, social power is impossible. This somewhat obvious condition tells us something important about social power: There is always a choice involved. With social power, "there is an element of voluntarism . . . the punishment could be chosen in preference to compliance" (Blau, 2003, p. 117). The relationship between Actor A's willingness to use force and Actor B's power is negative: The less likely Actor A is to use force to obtain social goods, the greater is the potential for social power for Actor B.

The last condition implies a consistent value hierarchy. If the value system of Actor A changes and she or he no longer values the social goods B controls, then there can be no power. If, on the other hand, Actor A consistently values the goods B has to offer, then social power will increase, if the other conditions hold as well. Try this thinking exercise: Think about the relationships that you have with various professors as exchange relations. Use each of these four variables—social capital, alternative, willingness to use force, and consistent value hierarchy—and ask the following: Who has more power and why? I believe you'll find that your professors have varying levels of power because of the different ways these factors align. At minimum, this implies that social power is not a simple function of bureaucratic position.

There are five possible responses to power. Four of the responses correspond to the four conditions of power and constitute attempts to change the balance of power. In other words, if you want to change the amount of power someone has over you, you could

- Increase your exchange capital by obtaining a good or service that the other person desires
- Find alternative sources to what you receive from the other person
- Get along without the good or service the other person controls
- Or, you could attempt to force the other person to give you what you need

The only other possible response is subordination and compliance. Compliance, unlike most features of social exchange, can be specified and functions like money: "Willingness to comply with another's demands is a generic social reward, since the power it gives him is a generalized means, parallel to money, which can be used to attain a variety of ends" (Blau, 2003, p. 22). The power to command is like a credit, an I.O.U. in social exchange. It is what we give to others when we can't participate in an equal exchange yet we desire the goods they control.

The *Web Byte* for this chapter—*Karen S. Cook: Power in Exchange Networks*—includes a different way of understanding power and exchange. Blau takes the exchange between two people (dyad) as the archetype of exchange. In contrast, Karen Cook focuses more on the network of exchange rather than the exchange itself. Social networks are defined by the patterns and positions of a group's interaction. For a quick example, think about an organizational chart. A person's position on that chart determines her or his general pattern of interaction. The same is true for any social group of which you are a member. Within a group there are specific patterns of interaction: Some members of the group interact more than others, and some members don't interact together at all, even though they are in the same group. Social networks, then, are another kind of structure that produces effects that don't originate with the individual, that don't come from social group membership or identity, and about which the individual may or may not be aware. Cook provides us with yet another way of understanding structure—check it out.

The Micro–Macro Link

Secondary exchange relations result from power and occur at the group level among those who are collectively indebted to someone or to another group. In order for secondary exchange relations to come into play, people who are individually indebted to a person or group must have physical proximity and be able to communicate with one another. For example, let's say you are tutoring a number of students in sociology without charging them money. Each of those individuals would be socially indebted to you, and each individual relationship would be subject to the dynamics of exchange (alternatives, marginal utilities, and norms). As long as you only met with each person individually and they were unaware of each other, the exchange relations would remain individual. On the other hand, if you decide that your time would be better spent tutoring them as a group, then secondary exchange relations could come into play. You've provided them with the ability to become a group through physical proximity and communication. When a group like this is brought together physically and able to communicate, two possibilities exist with regards to power: The group may either legitimate or oppose your power as the tutor.

Blau gives us a very basic process through which power is either legitimated or de-legitimated. Both possibilities are in response to the group's perception of how those in power perform with regard to the norms of fair exchange and reciprocity. If the norms of reciprocity and fair exchange are adhered to, then social power will be legitimated. Blau posits that the path looks something like this: reciprocity and fair exchange with those in power → common feelings of loyalty → norm of compliance → legitimation and authority → organization → institutionalized system of exchange values. As the group communicates with one another about the level of adherence of those in power to the norms of fair exchange and reciprocity, they collectively develop feelings of loyalty and indebtedness. Out of these feelings comes the norm of compliance: The group begins to sanction itself in terms of its relationship to those in power. This process varies in the sense that the more those in power are seen as consistently generous, that is, exceeding the norms of fairness and reciprocity, the more the group will feel loyal and the stronger will be the norm of compliance and sense of legitimation. Out of legitimation come organization and an institutionalized system of values regarding authority (see Blau & Meyer, 1987, for Blau's treatment of bureaucracy).

In his theory of secondary exchange relations, Blau is giving us an explanation of the micro–macro link. We can see this move from individual exchanges to institutions in the path of secondary relations noted above. The "glue" that holds this path together consists of generalized trust and the norm of reciprocity. As we've already seen, all social exchanges are built on trust; the element of time and lack of specificity in social exchanges demand trust. Organizations and institutions are, in Blau's scheme, long chains of indirect exchanges of rewards and costs. And, as we've seen, exchange intrinsically entails reciprocation. Every step, then, along the chain of exchanges is held together by the norm of reciprocity.

Secondary exchange relations can thus lead to legitimated authority, but they can also lead to opposition and conflict. If those in power do not meet the norms of fair exchange and reciprocity, and if those indebted are brought together physically and are able to communicate with one another, then feelings of resentment will tend to develop. These feelings of resentment obviously lead to de-legitimation of authority. This path of secondary exchange relations looks like this: lack of reciprocity and fairness from those in power → feelings of resentment → communication (a function of physical proximity and communication technologies) → de-legitimation of authority → ideology → solidarity → opposition → probability of change.

Groups experiencing a lack of fairness and reciprocity that are in close physical proximity and are able to communicate with one another will tend to develop a set of beliefs and ideas that both justify their resentment and their de-legitimation of authority. This ideology, in turn, enables group solidarity and overt opposition, thus increasing the probability of change. The last part of this path comes from the conflict theories of Marx and Weber. To this general theory of conflict and change, Blau adds the micro dimension of exchange: the beginning part of the path. Keep in mind that all of these factors function as variables and are therefore changeable.

Concepts and Theory: Population Structures

Dimensions of Population Structures

Population structures are only one of several macro-level influences on people's lives. Blau recognizes that historical trends impact not only people's lives but also the very structures with which he is concerned. He is also quite aware of the impact that cultural norms, values, and beliefs can have. In fact, his "focus on structural rather than cultural sociology is intended partly as a counterweight to the prevailing cultural emphasis in American sociology" (Blau, 1994, p. 16). Part of what we can see here is what Blau does *not* mean when he talks about social structures. He is not referring to institutions and practices such as capitalism and monogamy, nor is he referring to a country's stage in historical development, like the level of industrialization—all of which he grants have separate and important influences on society and individuals.

Blau defines *population structure* as the distribution of social positions in various dimensions. A social position is any kind of social category that is arranged hierarchically. Race, gender, religion, and class are all examples of social positions. Blau has a very specific idea about what makes these social positions structural. Structural properties do not refer to the content of the social positions nor to that which is average or common about the aggregates of subunits. The content of a social position refers to its unique characteristics, like what it means to be a woman as compared to a man in Western society. For Blau, these characteristics of the position are not structural. They may be important, but they aren't part of the social or population structure.

The demographic averages of the social positions aren't what Blau has in mind either. For example, we can say that the most common religious preference in Spain is Catholic and the most common in Norway is Lutheran. But neither of these statements is what Blau means when he talks about structure. Rather, structure refers to the *distribution of positions in structural space.* Another way to put this is that structure refers to the relationships of social positions when compared to one another. For example, the structural property of religion in both Spain and Norway is homogeneous when compared to the United States, which is heterogeneous.

Let's use the analogy of a building to understand Blau's idea. A building has walls, doors, a floor, and a ceiling. The relationship of all those parts one to another is what forms the structure of the building. Within the interior space of the structure, we can move about freely, choosing to go down one hall rather than another or to go in one room rather than another. The structure both enables and restricts our movements. We can think of the interior space as the actual content or meaning of social positions. For example, the interior space in this sense would be our race or religion. One may be, perhaps, Italian and Catholic. But these social positions in and of themselves don't constitute structure. Just as in the building, it is the relationships among the parts that creates structure. So, being Italian and Catholic in Italy is one thing, but it is entirely different to be those things in North Carolina. The structural properties, in this case, would be the relationships among various positions in Italy versus the southern United States.

There are three basic kinds of structural positions or parameters. Each of these parameters varies on a continuum. *Nominal parameters* refer to the distribution of a population among discrete categories or groups (like race, gender, and sexual preference). The extremes of the nominal continuum are heterogeneity and homogeneity, when *heterogeneity* is defined as the chance of two randomly chosen people belonging to different groups. The greater the number of social positions and the more evenly distributed the positions, the greater will be the level of heterogeneity within a population. For example, if we were to randomly pick two people in the United States and two people in Japan, the chances are good that the two people in the United States would be more diverse in terms of their social categories than the two from Japan, because the population in the United States has a greater number of social positions; its population is more heterogeneous in this respect when compared to Japan.

Graduated parameters refer to the distribution of a population on continuous rank orders (such as income, education, status, wealth, and power). The continuum extremes of graduated parameters are inequality and equality, where *inequality* is defined as the chance expectation of the absolute difference in given resources between two random people when compared to the mean resource difference in the population. The theoretical minimum of inequality is an even distribution of resources and the maximum is all resources held by one person. "In short, the more resources are concentrated in a few hands, the greater the inequality" (Blau, 1994, p. 14). The third structural parameter concerns the way these two other parameters come together. Positions on these parameters may be more or less correlated with one another and form *crosscutting social circles* (an idea that comes from Max Weber and Georg Simmel).

Crosscutting social circles captures the notion that different positions on the structural dimensions may or may not intersect with one another. Intersection refers to "the degree of independence of two or more attributes" (Blau, 2002, p. 350). So, for example, gender and annual salary in the United States are somewhat correlated with each other, but less so now than in the past. One hundred years ago, if you were a man and you met a woman, the chances were very high that her annual salary would've been less than yours. At that moment in history, those two dimensions intersected with one another in social space. Today, those two dimensions are more "consolidated" than they were previously.

Probabilities of Intergroup Relations

Blau (1994) characterizes this last parameter as "the most important emergent structural property" (p. 14). The reason for its importance is the way it interacts with various assumptions to produce the chances of intergroup relations (whether or not groups will come in contact with one another), which in turn affects the chances of social mobility and conflict. Let's look at how these things work out, and I think you'll see what Blau is talking about.

Blau begins his discussion of the probabilities of intergroup relations by stating two basic assumptions.

lower the number of intergroup associates, the lower will be the probability of mobility.

- *Theorem 7—the influence of heterogeneity on mobility:* If the level of heterogeneity is positively related to the probability of intergroup relations (Theorem 2), and if there is a positive relationship between number of intergroup relations and the probability of mobility (Theorem 6), then it follows that there is a positive relationship between heterogeneity and the probability of mobility.

- *Theorem 8—the influence of inequality on mobility:* If there is a positive relationship between inequality and probability of status-distant contacts (Theorem 3), and if there is a positive relationship between number of intergroup relations and the probability of mobility (Theorem 6), then there is a positive relationship between inequality and the probability of mobility. This may sound counterintuitive, but an illustration might help. For example, if Sue has associates that have higher status or occupational standings, then those associates can be of help in getting Sue a better job or a higher status position.

- *Theorem 9—the influence of group intersections on mobility:* If Theorem 5 and Theorem 6 are true, then there is a positive relationship between multiple intersections and the probability of mobility. In other words, if the various dimensions of social position are not strongly correlated, then individuals will have associates that belong to different groups or strata and those associations will tend to improve the person's mobility. For example, in the United States, race and occupation correlate—they tend to go together. If, however, occupation and race were separate and unrelated, then people of color would have a higher probability of meeting people of their race with high-paying jobs. This instance of crosscutting social circles (being similar on one dimension but different on another) would increase the probability of mobility for racial minorities.

- *Theorem 10—the influence of relative size on mobility:* If Theorem 1 and Theorem 6 are true, then there is an inverse relationship between group size and the probability of mobility (all other things being equal): proportionately smaller groups will experience higher levels of mobility than larger groups. For example, the proportion of elite downward mobility is greater than the proportion of the middle class that is upwardly mobile.

Population Structures and Conflict

Blau is interested in explaining interpersonal conflict, that is, conflict that involves direct contact between members of various groups. Thus, Blau is not considering such conflictual issues as ethnic discrimination or economic exploitation, both of which can occur apart from personal contact. It's also important for us to note that Blau is concerned with the structural influences on conflict that can be detected apart from substantive differences. For example, let's say that you and I disagree over the grade I gave you on the last test. This disagreement is a kind of conflict, but it is one that appears to be related to substantive issues (the grade dispute) more than structural issues. Blau's assumption is that in large populations, the probabilities of social relations and conflict can be seen even apart from purely individual or political issues.

In order to further contextualize what Blau is talking about with the structural determinates of conflict, let's remind ourselves that Blau has already given us a theory of conflict in his exchange relations. Because of the possibility of imbalanced exchanges, the potential for conflict is always present in exchanges. Imbalanced exchange brings power into the exchange relationship. But, as we've seen, the use of power must be exercised within the norms of reciprocity and fair exchange. If it is not, and if the group members can communicate with one another, then the group is likely to form an ideology justifying their perceptions and experience solidarity; and ideological solidarity facilitates conflict between social groups.

Blau's structural concerns, on the other hand, focus on the distribution of a population. In terms of the structural influences on conflict, Blau posits

- *Theorem 11—the influence of contact:* From Assumption 1, the probability of conflict depends upon social contact. Thus, the greater the probability of contact, the greater is the likelihood of conflict.
- *Theorem 12—the influence of relative size on conflict:* From Theorem 1, there is an inverse relationship between size and conflict, because smaller groups are more likely to have intergroup contact.
- *Theorem 13—the influence of heterogeneity on conflict:* From Theorem 1, Assumption 1, and the definition of heterogeneity: Heterogeneity has a positive relationship with intergroup conflict. The greater the level of heterogeneity in a population, the greater is the probability of intergroup conflict.
- *Theorem 14—the influence of status distance on conflict:* From Assumption 1 and the definition of inequality, inequality increases the probability that status differences will be involved in conflict.
- *Theorem 15—the influence of mobility on conflict:* From Assumption 1 and the definition of mobility, there is a positive relationship between mobility and intergroup conflict. Mobility increases intergroup conflict.
- *Theorem 16—the influence of consolidated versus intersecting social differences:* From Theorem 13 and Theorem 14, if various kinds of heterogeneity and inequality influence the level of conflict, then we can infer "that the consolidation of such differences further intensifies conflict between different groups and strata" (Blau, 1994, p. 41). Thus, if two groups experience parallel forms of inequality, heterogeneity, or both, then the conflict between them will be heightened. In other words, based on Theorem 16, we would expect tension and conflict between two minority groups suffering similar oppression to be exacerbated. On the other hand, "crosscutting cleavages keep conflicts within bounds. They are the result of multigroup affiliations and intersection in complex societies, which create cross-pressures for the many persons who belong to groups that have opposite views on some issues. . . . Such situations put individuals under cross-pressure, which may lead some not to take sides at all and undermine others' inflexible opposition convictions" (Blau, 1994, p. 41).

Summary

- Peter Blau works deductively, which means that his theory moves from abstract assumptions and definitions to logically derived theoretical statements (theorems). Blau focuses on two principle issues in his theorizing: social exchange and population structures. The reason for these two emphases is that they explain two of the three factors that influence human behavior (the other influence being psychological).

- Social exchanges are different from economic exchanges because they lack specificity, they require and build trust, they are meaningful, and social benefits are detached from the source. These differences imply that social exchanges create diffuse obligations, which in turn form the basis of society—social relations must be maintained in order to guarantee repayment of these obligations. Population structures are made up from the distributions of a population along various continuums of difference or social position. These continuums of difference create or hinder opportunities for social contact, social mobility, and social conflict.

- There are three basic principles and two norms of exchange. The contours of all social exchanges are set by the principles of rational motivation, the presence of alternatives, and satiation. Because of the peculiar properties of social exchanges, rationality within them is limited, as compared to economic exchanges. People are rational in social exchanges to the extent that they tend to repeat those actions that they received rewards from in the past. People are also rational in the sense that they will gravitate toward exchanges that are equal, the equality of exchanges being determined by the presence of alternatives. And all social exchanges are subject to the principle of marginal utilities—a social good loses its value in exchange as people become satiated; in other words, value in exchange is determined to some extent by scarcity and uncertainty. All social exchanges are subject to the norms of reciprocity and fair exchange.

- Social actors achieve power through unequal exchanges, with inequality in exchange determined by four factors: the level of exchange capital, the number of potential source alternatives, the willingness and ability to use force, and consistency in value hierarchy. The first three are negatively related to power. That is, in an exchange relationship between Actor A and Actor B, as Actor A's capital, alternatives, and ability to use force go down, Actor B's power over Actor A increases. Consistency is a positive or at least steady relationship—continuing to value the goods that Actor B controls places Actor A in a possible subordinate position.

- The presence of power sets up the possibility of secondary exchange relations. These occur at the group level between a supplier and consumer, where the consumer is a group of people who rely upon a single supplier. Under such conditions, the power of the supplier will either be opposed or legitimated. If the norms of reciprocity and fair exchange are adhered to, then the power of the supplier will be legitimated and feelings of loyalty and a norm of compliance will emerge. From this base, organizations and institutionalized systems of exchange are built.

- There are three dimensions to population structures: nominal parameters, graduated parameters, and the relationship between nominal and graduated parameters. Nominal parameters are made up of the discrete social categories used by a population (such as race and gender). Graduated parameters are those distributions of a population on continuous rank orders (like income). Positions on nominal and graduated scales may be more or less correlated with one another (as when race and income level go together).

- There are five structural issues that influence the likelihood of intergroup contact (the possibility of contact between members of different groups). The likelihood of intergroup contact is a negative function of relative group size, and a positive function of population heterogeneity, inequality, mobility (broadly defined), and the intersection of social circles.

- There are five structural features that affect the level of mobility. Mobility is broadly defined in terms of any movement. Mobility is a positive function of the number of intergroup relations, population heterogeneity, social inequality, and group intersections. Relative group size and mobility are inversely related.

- Blau presents a theory explaining the structural determinates of conflict. These are influences on conflict that exist apart from any political or personal issues. Blau's theory is concerned with explaining the probability of interpersonal conflict that involves the direct contact of members of different groups. There are six structural factors that influence the probability of interpersonal conflict. The likelihood of interpersonal conflict is a positive function of the extent of contact, heterogeneity, status distance, and mobility. Interpersonal conflict between group members is a negative function of the relative size of the group. In addition, consolidated structural dimensions will tend to increase conflict, while intersecting social differences will tend to decrease it.

Building Your Theory Toolbox

Learning More—Primary Sources

- Blau has written two short pieces that provide an excellent overview of his ideas:
  - Social exchange. In David L. Sills (Ed.), *International encyclopedia of the social sciences*, Macmillan, 1968.
  - Macrostructural theory. In Jonathan H. Turner (Ed.), *Handbook of sociological theory*, Kluwer Academic/Plenum, 2002.
- For more complete explanations of these issues, see the following works by Peter Blau:
  - *Exchange and power in social life*, Transaction Publishers, 2003.
  - *Structural contexts of opportunities*, University of Chicago Press, 1994.
  - *Crosscutting social circles: Testing a macrostructural theory of intergroup relations*, Academic Press, 1984.

Learning More—Secondary Sources

- *Structures of power and constraint: Papers in honor of Peter M. Blau,* edited by Craig J. Calhoun, Marshall W. Meyer, and W. Richard Scott, Cambridge University Press, 1990.

Check It Out

- *Web Byte— Karen S. Cook: Power in Exchange Networks*
- *Utilitarianism* is a philosophy of morals and government that has had a strong influence on the formation of the United States. The first two are primary texts and the third is a contemporary look at the implications:
  - o Jeremy Bentham: *An introduction to the principles of morals and legislation,* Oxford University Press, 1966.
  - o John Stuart Mill: *Utilitarianism,* Hackett, 2002.
  - o William Shaw: *Contemporary ethics: Taking account of utilitarianism,* Blackwell, 1998.

Seeing the World

- After reading and understanding this chapter, you should be able to answer the following questions (remember to answer them *theoretically*):
  - o What are the unique features of social exchange as compared to economic exchange?
  - o Explain what power is, from Blau's perspective, and how it comes out of social exchanges.
  - o What are the norms of reciprocity and fair exchange? How are they important?
  - o What are secondary exchange relations and how do they account for the micro–macro link?
  - o What are population structures and how are they different from what we usually call social structures? In your answer, be sure to explain the three dimensions of population structures.
  - o Using Blau, explain the factors in back of a high probability of conflict in any given society.

Engaging the World

- One of the enlightening aspects of exchange theory is that it helps us understand our relationships in terms of exchange. For example, what kinds of exchange dynamics are at work in your family or significant relationships? How can you understand your relationship with your professor using exchange theory? Specifically, how is power achieved and how could you reduce the level of power?
- I'd like you to look at your school in terms of its population structures. Because of the kind of society we live in, many of the nominal parameters that make up your school's population structure are readily available. All universities are required to collect such data. At the university where I work, the data are accessible through the Office of Institutional Research

under "Student Data Profiles." Your university will have a similar office, and the information is undoubtedly available through your campus Web site. You can either access the graduated parameters or you can use information that is available to make educated guesses. Using this information and Blau's theory, predict and explain the probability of intergroup contact.

Weaving the Threads

- How can you understand SI's notion of interaction using Blau's theory of the probability of social contact? Can the two theories be joined? How would Goffman's ideas of rounds and settings fit in? These kinds of questions get at the fundamental issue of how behaviors and interactions are patterned.
- Garfinkel gives us a theory of how social order is achieved at the micro level. Blau gives us a theory of how power is achieved and maintained at the micro level. How can these two theories be joined? Is power a part of social order? If so, how?
- Compare and contrast Blau's and Collins's idea of power. What is power and how does it work in these two theories?
- Compare and contrast Blau's and Collins's theories of the micro–macro link. Put them together in such a way as to begin to form a single theory. What do you think is still missing?

CHAPTER 7

Gender Inequality

Janet Saltzman Chafetz (1942–)

Photo: Courtesy of Henry Chafetz.

Throughout history, the social category of gender has been and continues to be the most fundamental way in which distinctions are made among people. As such, gender as a social category influences almost every form of discrimination. Take the obvious example of class. According to the Business and Professional Women's Foundation (2004), women generally get paid 76% of what a man makes for the same job with the same education, which means that women are underpaid by about half a million dollars over a lifetime. Interestingly, this discrepancy increases for some of the best-paying and most powerful occupations. Female physicians, for instance, earn about 68% of what comparable male doctors do.

Women also experience what is referred to as "the glass ceiling," an invisible wall that stops women from advancing up corporate and political ladders past a certain point. Thus, men hold most of the top company positions in the United States: In 2002, there were only six female chief executive officers in the Fortune 500. Men also hold most of the political positions: In the 108th Congress (2003–2004), women held only 14% of the seats in both the U.S. Senate and the House of Representatives.

The list of the social consequences of gender is almost endless: Women are more likely to have been sexually molested than men; women suffer far more domestic violence than men; women have less decision-making power in organizations and relationships; women typically receive less education than men; women control fewer conversations than men; women are more supportive in conversations than men; the sexual expectations for women are oppressive when compared to men; women are more likely to be hassled in public than are men; and on and on.

In Western European societies, these inequalities of gender have been addressed through at least two waves of feminism. Inklings of the first wave began when ideas about the equal rights of women emerged during the Enlightenment. The first significant expression of these concerns was Mary Wollstonecraft's book *A Vindication of the Rights of Woman* (1792). But the first wave of feminism didn't become organized until the 1848 Seneca Falls Convention, which called for equal rights to vote and own property, full access to educational opportunities, and equal compensation for equal work.

The second wave of feminism grew out of the civil rights movements of the 1960s. Publication of Simone de Beauvoir's *The Second Sex* (1949) and Betty Friedan's *The Feminine Mystique* (1963) was particularly important for this second wave of feminists, as was the founding in 1966 of the National Organization for Women (NOW). Central issues for this movement were pay equity; equal access to jobs and higher education; and women's control over their own bodies, including but not limited to sexuality, reproduction, and the eradication of physical abuse and rape.

In the early 1970s, two clear divisions began to appear among second-wave feminists. One camp emphasized the more traditional concerns of women's rights groups and generally focused on the similarities between men and women. Their primary concern was structural inequality. The other group moved to more radical issues. Rather than emphasizing similarities, they focused on fundamental differences between men and women. In some ways, the concerns of this group, described in the paragraph below, are more radical than equal opportunity. This more critical group is often described as a "third wave" of feminism.

The idea of third-wave feminism began to take hold around the intersection between race and gender—there are marked distinctions between the experiences of black and white women. But more recently, it has gained currency with reference to age. It appears that the experiences of young, contemporary feminists are different from those of second-wave feminists. Young feminists grew up in a social world where feminism was part of common culture. These young women are also playing out some of the postmodern ideas of fluid identities. The result is that many young feminists can best be described through contradiction and ambiguity. As Jennifer Drake (1997) says in a review essay, "What unites the Third Wave is our negotiation of contradiction, our rejection of dogma, our need to say 'both/and'" (p. 104). For example, third wavers might claim their right to dress sexy for fun while simultaneously criticizing patriarchy for objectifying women.

In this book, there are three theorists that specifically address gender. Two of these theorists, Dorothy Smith and Judith Butler, are among those that take the more radical view of feminism. For them, the life experiences and perspectives of men and women are different, and gender equality is more fundamental than structural equality. Dorothy Smith (Chapter 17) focuses on the unique consciousness and lived experience of women. Smith argues that women have a bifurcated awareness of the world, split between the objective world of men and their own lived experiences. Judith Butler (Chapter 18) focuses first on the body and secondarily on consciousness, or more accurately, subjectivity. She argues that gender inequality presumes a more basic construction: the normative sexing of the body.

Both Smith and Butler present critical understandings that show how women are controlled through systems of knowledge and embodiment. This type of control is deeply embedded in the mind and body, and it generally functions below the level of awareness. The key to sex or gender equality for Smith and Butler seems to lie in subversion—practices that critically expose the assumptions upon which the control of women is based.

Janet Saltzman Chafetz, on the other hand, presents us with a way of understanding gender that is more in keeping with the structural issues raised by both the first and second waves of feminism and is more closely aligned with mainstream sociology and positivistic theorizing. Chafetz synthesizes theories, just as Smith and Butler do, but she generally draws on less critical or radical perspectives, such as Marxian stratification and exchange theories. Chafetz also approaches theory more conventionally from a positivistic point of view. She creates propositions and dynamic models that explain and predict gender stratification. For example, in what is perhaps her most influential book, *Gender Equity*, she presents no fewer than 21 models and 64 propositions. Chafetz's central concern is also different: Rather than the control of women through consciousness or embodiment, she is specifically concerned with the unequal distribution of scarce resources.

One thing is certain: Gender is a complex issue, and our three theorists give us three inroads to understanding sex and gender. Chafetz addresses the social structure, Smith speaks to consciousness, and Butler gives us insights into the sexed body. If we can accept the idea that society, social relations, and social beings are made up of different things that respond to diverse theories and explanations, then we can think of Chafetz as providing the broad, structural basis for gender inequality and Smith and Butler as explaining some of the most important yet subtle ways in which women are actually governed.

The Essential Chafetz

Biography

Janet Saltzman Chafetz was born in Montclair, New Jersey, in 1942. She received her B.A. in history from Cornell University and her M.A. in history from the University of Connecticut. While at the University of Connecticut, she began graduate studies in sociology and completed her Ph.D. at the University of Texas at Austin. Chafetz served as president of Sociologists for Women in Society (SWS), 1984–1986, and as chairperson of the American Sociological Society (ASA) Theory Section, 1998–1999. Chafetz was also honored as the first invited lecturer for the Cheryl Allyn Miller Endowed Lectureship Series, sponsored by SWS; her book *Gender Equity* won the American Educational Studies Association Critic's Choice Panel Award (1990) and was selected by *Choice Magazine* for their list, Outstanding Academic Books (1990–1991). Chafetz is currently Professor of Sociology at the University of Houston and has recently been working on immigrant and transnational religion.

Passionate Curiosity

Chafetz is a positivist seeking to understand and facilitate change in the system of gender inequality. She approaches gender inequality like a scientist would approach cancer. In order to eradicate cancer, the laboratory researcher first has to understand how it works. Understanding how it works can lead to targeted methods of cure rather than trial-and-error shots in the dark. As Chafetz (1990) says, "In practical terms, a better understanding of how change occurs . . . could contribute to the development by activists of better strategies to produce change" (p. 100).

Keys to Knowing

unpaid labor force, women's workforce participation, organizational personality, social exchange and micro power, gender definitions, intrapsychic structures, engenderment, unintentional and intentional gender change

Normally I separate out a theorist's perspective and her or his theory. With Chafetz, part of her perspective has already been covered above: She is a feminist who uses sociological theory to explain and reduce structured gender inequality. But we are going to have to cover the remainder of her perspective as we explain her theory. One of the delights in working with someone like Chafetz is the amount and diversity of theoretical perspectives and principles she brings together. In explaining gender inequality, Chafetz brings together the work of Karl Marx, Rosabeth Kanter, Nancy Chodorow, Albert Bandura, Erving Goffman, and Ralf Dahrendorf, as well as elements from exchange theory and social movements theory. So, with Chafetz, it's much easier for all of us to review her perspective as we go through her theory. The powerful thing about what Chafetz does is that she makes all those diverse theories fairly easy to understand.

To begin, Chafetz assumes gender inequality. She isn't so much interested in explaining how gender inequality came about as she is in explicating how gender systems of inequality are stabilized and maintained. And she is quite inclusive in her explanation. She considers all three levels of analysis—macro, meso, and micro—and she argues that gender stratification is maintained through voluntaristic as well as coercive means. Chafetz argues that coercion usually functions as a background feature that helps to legitimate gender inequality. Both men and women generally conform to gender expectations and coercion isn't necessary. Coercion usually only comes to the fore during times of change or uncertainty.

Concepts and Theory: Coercive Gender Structures

Macro-Level Coercive Structures

Chafetz draws from specific theories to explain how the coercive and voluntaristic features work. The coercive theories tend to correspond to the three levels of analysis, so that there is a particular theory that is used at each level. Her primary orientation for the macro-level structural features of gender stratification comes from Marxian feminist theory. The basic orientation here is that patriarchy and capitalism work together to maintain the oppression of women. And the central dynamic in the theory is women's workforce participation.

Marx argued that social structure sets the conditions of social intercourse. In this sense, society itself has an objective existence and it thus strongly influences human behavior. More than that, social change occurs because of structural change. In Marx's theory, the economic structure itself contains dynamics that push history along. In general, Chafetz draws from Marx's emphasis on the economy as the most important site for social stability and change. She also explores his ideas about the way capitalism works.

Capitalism requires a group who controls the means of production as well as a group that is exploited. This basic social relationship is what allows capitalists to create profit. Patriarchy provides both: men who control the means of production and profit and women who provide cheap and often free labor. That latter part is particularly important. Much of what women do in our society is done for free. No wages are paid for the wife's domestic labor—this work constitutes the *unpaid labor force* of capitalism. Without this labor, capitalism would crumble. Paying women for caring for children and domestic work would significantly reduce profit margins and the capitalists' ability to accumulate capital. In addition, the man's ability to fully work is dependent upon the woman's *exploitation as a woman* as well. When women are allowed in the workforce, they tend to be kept in menial positions or given lower wages for the same work as men.

Because of the importance of women's cheap and free labor to the capitalist system, elite males formulate and preach a patriarchic ideology that gives society a basis for believing in the rightness of women's primary call to childrearing and domestic labor. Elite men also use their structural power to disadvantage women's workforce participation. For example, in the United States, elite males have been able to systematically block most attempts at passing national comparable worth amendments or laws that would guarantee equal pay for equal work.

In brief, Chafetz argues that the greater the *workforce participation* of women, particularly in high-paying jobs, the less the structure of inequality is able to be maintained. The inverse is true as well: the less workforce involvement on the part of women, the greater the inequality on all levels. Thus, the type and level of their involvement in the workforce (macro) play out at both the meso and micro levels. Though we've framed our discussion in terms of capitalism, Chafetz notes that since humans began to farm and herd animals, men have disproportionately controlled the means of production and its surplus. Throughout time, men have been slow to give up their economic positions.

Meso-Level Coercive Structures

To explicate the dynamics that sustain gender inequality at the meso level, Chafetz cites Rosabeth Kanter's (1977) work on organizations. Kanter gives us a social-psychological argument where the structural position of the person influences her or his psychological states and behaviors. Kanter points to three factors related to occupational position that influence work and gender in this way: the possibility of advancement, the power to achieve goals, and the relative number of a specific type of person within the position. Each of these factors in turn influences the individual's attitudes and work performance—or what we could call her or his *organizational personality.*

The Possibility of Advancement

Most positions in an organization fall within a specific career path for advancement. The path for a professor, for example, goes from assistant, to associate, to full professor. The position of dean doesn't fall within that path. A professor could aspire to become dean, but she would have to change her career trajectory. Kanter argues that women typically occupy positions within an organization that have limited paths for advancement. We can think of the occupational path for women as constricted in two ways: (1) the opportunities for advancement in feminized occupations, such as administrative assistant or secretary, are limited by the nature of the position; and (2) women who are on a professional career path more often than not run into a glass ceiling that hinders their progress.

The Power to Achieve Goals

Positions also have different levels of power associated with them. Again, this is a feature of the location within the organization, not of the individual. For example, if you were in my theory class, I would have you write theory journals. Most of my students do the journals because I tell them to. But my authority to assign work doesn't have anything to do with me; it's a quality of the position. If there were another person in my position, the students would do what she or he told them. And, again, women typically hold positions with less power attached to them than do men. Clearly, there are some women who transcend this situation. But women that are in positions of power are typically seen as "tokens," because there are no similar others in those positions within the company (Kanter's third organizational variable).

Relative Numbers ·

One of the things within organizations that facilitates upward mobility is the relative number of a social type within a position. Imagine being a white male and reporting to work on your first day. You're given a tour of the facilities. As you are introduced to different people in the company, you notice that almost all the positions of power are held by black men. Most of the offices are occupied by black men, and most of the important decisions are made by black men. There are whites at this place of employment, but they almost all hold menial jobs. By and large they are the secretaries and assistants and frontline workers. Being white, how would you gauge your chances for advancement at such a firm?

Further, imagine that after the end of your first week, you notice that there are two or three whites who seem to hold important positions. Would the presence of a few whites in management change your perception? It isn't likely. Though we like to talk as if individuals are the only things that matter, in fact, humans respond more readily to social types than to individual figures. That's why the few minorities that do make it up the corporate ladder are seen as tokens—exceptions to the rule—rather than as any real hope that things are changing.

As I've pointed out, these qualities are attributes of the position more than the person. This issue is important for Kanter's argument because social contexts influence individuals and their attitudes and behaviors. Positions that are similar in the organization produce similar contexts for people that most importantly include power and opportunity. These contexts influence the way the incumbents—the people that occupy the position—think and act.

In situations where power is available and the gates in the organizational flow lines are open, people develop a sense of efficacy. They feel empowered to control their destiny within the organization, and they behave in a "take charge" manner. The reverse is true as well. People in positions where there are few opportunities and where power is limited are much less certain about showing positively aggressive behaviors. They feel ineffectual and limited in what they can achieve within the organization.

These issues are true for any who occupy these different positions. Humans are intimately connected to their context. Our social environment always influences who we are and how we act. The problem in organizations is that women are systematically excluded from positions of power and opportunity. As a result, they experience and manifest a self that corresponds to the position, one that feels and behaves ineffectual and limited. Though these differences in behavior and attitude are linked to organizational position, people generally attribute them to the person. Thus, women who occupy positions that have little power or hope of advancement demonstrate powerlessness and passivity. These attitudes and behaviors are then used to reinforce negative stereotypes of gender and work, which, in turn, are used to reinforce gender inequality within the organization.

Micro-Level Coercive Structures

Before we begin this section, I want to point out that even though we are talking about the micro level, Chafetz is talking about structures that have the power of coercion. This is important to keep in mind because many micro-level theories are oriented toward choice and agency. Chafetz talks about those issues in the section on voluntaristic gender inequality, but for now we are considering how gender inequality is a structural force at the micro level.

Chafetz uses exchange theory to explicate coercive processes at the micro level (see Chapter 6). Exchange theory argues that people gravitate toward equal exchanges. Both partners in an exchange need to feel they are getting as much as they are giving. If an exchange isn't balanced, if one of the participants has more resources than the other, the person who has less will balance the exchange by offering compliance and deference. According to exchange theory, this is the source of power in social relationships.

Exchange theory also makes a distinction between economic and social exchanges. Economic exchanges are governed by explicit agreements, often in the form of contracts. The particulars of the exchange are well known in advance, and there is a discernable end to the exchange. For example, if you are buying a car on credit, you know exactly how much you have to pay every month and when the payments will stop. You know when your debt is paid off. However, in social exchanges the terms of the exchange cannot be clearly stated or given in advance. Imagine a situation where a friend of yours invites you over for dinner and tells you when and how you will be expected to repay. Chances are you wouldn't go to dinner because the other individual broke the norms that make an exchange social. Thus, social exchange is implicit rather than explicit and it is never clear when a debt has been paid in full.

Chafetz argues that these two issues together create a coercive micro structure that perpetuates gender inequality for women. Because of their systematic exclusion from specific workforce participation, women typically come into intimate relationships with fewer resources than men in terms of power, status, and class. The imbalance is offset by the woman offering deference and compliance to the man. This arrangement gives the man power within the relationship. The man's power in this exchange relationship is insidious precisely because it is based on social exchange. As we've seen, social exchanges are characterized by implicit agreements rather than explicit ones, with no clear payoff date or marker.

While insidious, this *micro power* is variable. Generally speaking, the more the economic structure favors men in the division of labor, the greater will be a man's material resources relative to the woman's, and the greater will be his micro-level power. The inverse is also true: "The higher the ratio of women's material resource contribution to men's, the less the deference/compliance of wives to their husbands" (Chafetz, 1990, p. 48).

The greater power that men typically have is used in a variety of ways. One common way is in relation to household work. Much of the work around the home, especially in caring for the young, is dull, repetitive, and dirty. Men typically choose the kinds of tasks that they will do around the home, as well as the level of work they contribute. Thus, men usually do more of the occasional work rather than repetitive work, such as mowing the lawn or fixing the car rather than the daily tasks of doing dishes or changing diapers. Men can also use their power to decide whether or not women work out of the home and to influence what kinds of occupations their wives take. Because women in unbalanced resource relations bear the greater workload responsibility for the children and home, they are restricted to jobs that can provide flexible hours and close proximity to home and school.

Concepts and Theory: Voluntaristic Gender Inequality

As we noted before, gender inequality usually functions without coercion. This implies that women cooperate in their own oppression. More exactly, "people of both genders tend to make choices that conform to the dictates of the gender system status quo" (Chafetz, 1990, p. 64). Chafetz refers to this as voluntaristic action;

but, as we'll see, some of these behaviors and attitudes are unthinkingly expressed and so aren't "voluntary" in the sense of chosen. The patterns that support gender inequality are latently maintained; they are quiet and hidden.

The reason for this kind of maintenance is simple, and it is how society works in general: We simply believe in the culture that supports our social structures. Max Weber pointed this out many years ago. Every structural system is sustained through legitimation, whether it is a system of inequality or the most egalitarian organization imaginable, and *legitimation provides the moral basis for power*—it gives us reasons to believe in the right to rule. Part of the way legitimation works has to do with the place culture has in human existence: Culture works for humans as instinct does for animals. Another important reason for the significance of legitimation is that coercive power is simply too expensive, in terms of costs of surveillance and enforcement, to use on anything but an occasional basis.

Thus, Chafetz argues that much of what sustains the system of gender inequality is voluntarism. Both men and women continue to freely make choices and display behaviors that are stereotypically gendered. There are three types of *gender definitions* that go into creating gendered voluntaristic action: gender ideology, norms, and stereotypes. These three types vary by the level of social consensus and the extent to which gender differences are assumed.

Chafetz draws on three theoretical traditions to explain these issues: Freudian psychodynamic theory; social learning theory; and theories of everyday life, including symbolic interactionism, ethnomethodology, and dramaturgy. It's interesting to note that Chafetz is still working with levels of analysis: Psychodynamic theory addresses the inner structural core of the individual, social learning focuses on the behaviors of the person, and theories of everyday life explicate the interaction.

Intrapsychic Structures

Nancy Chodorow's work (1978) forms the basis of Chafetz's psychodynamic theory. Chodorow argues that men's and women's psyches are structured differently due to dissimilar childhood experiences. The principle difference is that the majority of parenting is given by the mother with an absent father. Before we talk too much about Chodorow's theory, we need to make sure we understand the idea of psychic structure, or, more specifically, intrapsychic structure.

Freud argues that the psychic energy of an individual gets divided up into three parts: the id, the ego, and the superego. These three areas exist as structures in the obdurate sense. They are hard and inflexible and produce boundaries between the different internal elements of the person. Thus, when we are talking about *intrapsychic structures,* we mean the parts of the inner person that are fixed and divided off from one another. It's almost like we're talking about the structure of the brain. The "intra" part of intrapsychic structures refers to how these three parts are internally related to each other. Thus, Chodorow's, and Chafetz's, argument is that the internal workings of boys and girls are structurally different. However, the difference isn't present at birth but is produced through social experiences.

Both boys and girls grow up with their chief emotional attachment being with their mother. Girls are able to learn their gender identity from their mothers, but boys have to sever their emotional attachment to their mother in order to learn their gender identity. The problem, of course, is that the father has historically been absent. His principle orientation is to work, a situation that was exacerbated through industrialization and the shift of work from the agricultural home to the factory.

Girls' intrapsychic structure, then, is one that is built around consistency and relatedness—they don't have to break away to learn gender and their social network is organically based in their mother. Women, then, value relationships and are intrapyschically oriented toward feelings, caring, and nurturing. Boys have to separate from their mother in order to learn gender, but there isn't a clear model for them to attach to and emulate. More exactly, the model they have is "absence."

The male psyche, then, is one that is disconnected from others, values and understands individuality, is more comfortable with objective things than relational emotions, and has and values strong ego boundaries. According to Freud, this male psyche also develops a fear and hatred of women (misogyny)—as the boy tries to break away from his mother, she continues to parent because of the absent father. The boy unconsciously perceives her continued efforts to "mother" as attempts to smother his masculinity under an avalanche of femininity. He thus feels threatened and fights back against all that is feminine.

Remember, these differences between males and females are dissimilarities in the structure of their psyches. This is a much stronger statement than saying that boys and girls are socialized differently. Intrapsychic structures are at the core of each person, according to Freudian theory, and much of what happens at this level is unconscious. Chafetz isn't necessarily saying that men consciously fear or hate women—it's much deeper than that. It is at the core of their being. But also keep in mind that these structures vary according to the kind of parenting configuration a child has. It's very possible today for a boy to be raised principally by the father while the mother works (though our economic structure makes this unlikely), or for a single father to raise a child. The intrapsychic structure of such a boy would be dramatically different than those raised in a situation where maleness is defined by absence and separation.

These intrapsychic differences play themselves out not only in male/female relationships, but also in the kinds of jobs men and women are drawn to. Generally, then, men are drawn to the kinds of positions that demand individualism, objectification, and control. Women, on the other hand, are drawn to helping occupations where they can nurture and support. While there have certainly been changes in occupational distribution over the past 30 years, most of the stereotypically gendered fields continue to have disproportionate representation. Thus, for example, while there are more male primary school teachers and nurses today than 30 years ago, the majority continue to be women, and while there are more women who are CEOs, construction workers, and politicians today, those fields continue to be dominated by men. The important point that Chafetz is bringing out here is that the "personal preference" individuals feel to be in one kind of occupation rather than another is strongly informed by gendered intrapsychic structures. These personal choices, then, "voluntarily" perpetuate gender structures of inequality.

Social Learning

Chafetz also draws on socialization theories such as social learning to explain the voluntaristic choices men and women make. The important components of social learning theory come to us from Albert Bandura (1977). Bandura argues that learning through experimentation is costly and therefore people tend to learn through modeling. For example, it's much easier to learn that fire is hot by the way others act around it than by sticking your hand in it. Social learning occurs through four stages: attention, retention, motor reproduction, and motivation.

Children pay attention to those models of behavior that seem to be the most distinctive, prevalent, or emotionally invested or have functional value. And they retain those models through symbolic encoding and cognitive organization, as well as rehearsing the behaviors symbolically and physically. Motor reproduction refers to acting out the behaviors in front of others. Further motivation to repeat behaviors comes through rewards and positive reinforcement. Children are discouraged from repeating inappropriate behaviors through punishment and negative reinforcement. Eventually, children negatively or positively reinforce their own behaviors—as adults we thus self-sanction most of our gendered behaviors.

Performing Gender

Theories of everyday life look at how people produce social order at the level of the interaction and manage their self-identities. Chafetz specifically draws on Erving Goffman's (1977) work on gender. Goffman argues that selves are hidden and the only way others know the kind of self we are claiming is by the cues we send out. Others read these cues, attribute the kind of self that is claimed, and then form righteously imputed expectations. People expect us to live up to the social self we claim. If we don't, that part of our self will be discredited and stigmatized.

Because gender is arguably the very first categorization that we make of people—one of the first things we "see" about someone is whether the person claims to be male or female—gender is thus an extremely important part of impression management and self-validation. As such, Goffman (1959) would see gender as a form of idealized performance in that we "incorporate and exemplify the officially accredited values of the society" (p. 35). In other words, social norms, ideologies, and stereotypes are used more strongly in gendered performances than in most others. Goffman also sees idealization as referencing a part of the "sacred center of the common values of the society" (p. 36) and this kind of performance as a ritual.

Thus, gender is an especially meaningful and risky performance that is produced in almost every situation. We tend to pay particular attention to the cues we give out about our gender and the cues others present. Goffman further points out that we look to the opposite gender to affirm our managed impression. Our gendered performances then are specifically targeted to the opposite sex and tend to be highly stereotypical. According to Chafetz (1990), "For men, this quest [for affirmation] entails demonstrations of strength and competence. However, for women it entails demonstrations of weakness, vulnerability, and ineptitude" (p. 26).

Concepts and Theory: Stratification Stability

In Figure 7.1, we have Chafetz's model of gender stability. In it we can see all of the theoretical issues we've discussed coming together. Chafetz argues that gender inequality is initially stabilized through the economic structure and the division of labor. A gendered division of labor gives men and women different resource levels. Beginning with the top part of the model, we can see that because men have superior material resources, women at the micro level must balance the exchange by offering deference and compliance. Wifely compliance means that women are either kept at home, and thus utterly dependent upon the man for material survival, or are allowed to work but must also carry the bulk of the domestic duties (double workday). Wives' absence or double workday feeds back to and reinforces the gendered economic division of labor.

Follow the other path coming from male micro-resource power to male micro-definitional power, which, in turn, leads to gender social definitions. Here Chafetz is arguing that men have the greatest influence on society's definitions of male and female due to their greater control of resources. These definitions include stereotypes, norms, and ideologies that devalue women's work, legitimate unequal

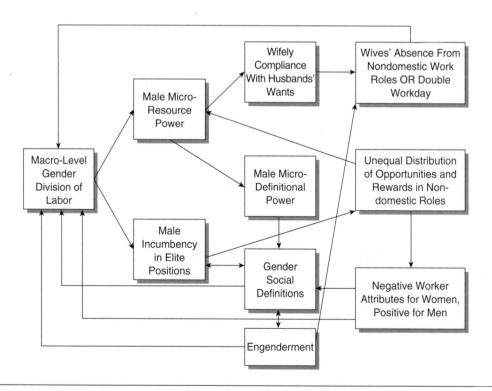

Figure 7.1   Gender Stratification Stability

SOURCE: From Chafetz, J. S., *Gender Equity: An Integrated Theory of Stability and Change*, copyright © 1990, Sage Publications, Inc. Reprinted with permission.

opportunities and rewards, and place higher value on masculine traits rather than feminine (Chafetz, 1990, pp. 84–85).

This masculinized culture produces engenderment. **Engenderment** is Chafetz's term to talk about the internalization of gendered selves, identities, and behaviors. It specifically includes psychodynamic as well as social learning processes. Notice that gender social definitions are influenced by male incumbency in elite positions. The incumbency variable comes from the meso-level theories of organization. Thus, social definitions of gender are produced by male superior power at both the micro and meso levels.

As you can see from the model, engenderment feeds into wives' absence from the economy and the double workday. Chafetz (1990) is telling us that "a major component—conscious or not—of the feminine personality and self-identity is an orientation toward nurturant and caretaking roles . . ." and that "women, therefore, choose to place priority on family responsibilities, and, where financially possible, this priority often includes the choice to forgo other forms of work altogether" (p. 75). Engenderment also reciprocally influences the macro-level division of labor and definitions of gender.

The other main effect coming from the macro division of labor is male incumbency. The elite of any group are more likely to give positions of power to group members than to outsiders. This tendency toward like members is particularly strong during times of uncertainty or threat. In addition, given the current social definitions of gender, "elites are likely to believe that women lack the personal traits required to fill positions of responsibility" (Chafetz, 1990, p. 53). Taken together, the initial gender inequality in the division of labor with men holding the positions of greater power plus gendered stereotypes of work expectations imply that there will be a strong tendency for men to guard and keep the positions of power for other men. This incumbency, in turn, creates an unequal distribution of opportunities and rewards, which helps create male resource supremacy at the micro level as well as producing negative worker attributes for women and positive ones for men at the meso level of organizations. Note that though not indicated on the model, all these relationships are positive (vary in the same direction) and thus work to structurally maintain gender stratification.

Concepts and Theory: Gender Change

To begin this section, we should note that Chafetz argues that structural rather than cultural changes are necessary to bring about gender equality. As we will see throughout this book, sociologists give different weights to culture and structure. Some argue that culture is an extremely important and independent variable within society; others claim that culture simply reinforces structure and that structure is the most important feature of society. Chafetz falls into the latter camp. While voluntaristic processes, which are associated in one way or another with culture, are the key way gender inequality is sustained, "substantial and lasting change must flow 'downward' from the macro to the micro levels" (Chafetz, 1990, p. 108).

Unintentional Change

Like her understanding of gender stability, Chafetz divides her theory of gender change into unintentional and intentional processes. Quite a bit of the change regarding the roles of women in society has been the result of unintended consequences. For example, the moves from hunter-gather to horticulture to agrarian economies were motivated by advances in knowledge and technology. But these moves also produced the first forms of gender inequality, as men came to protect the land (which gave them weapons and power) and to control economic surplus through inheritance and the control of sexual reproduction (women's bodies). The intent behind technological advancement wasn't gender inequality; the intent was first survival and then to make life less burdensome. But in the end, economic developments produced gender inequality.

In explicating the unintended change processes, Chafetz is interested in what she terms the *demand side.* She argues that quite a bit of work on gender focuses on the *supply side,* which concentrates on the general attributes of women. For example, contemporary women tend to have fewer children, to be better educated, and to marry later than in previous generations. These supply side attributes may influence what kinds of women become involved in the economy, but they "do not determine the rate of women's participation" (Chafetz, 1990, p. 122). For example, a woman may have a master's level education, but unless there is a structural demand for this quality, she will remain unemployed.

In addition, Chafetz assumes gender stability in theorizing about change. This means that males are the default to occupy a given position in the economy. The bottom line here is that *"as long as there are a sufficient number of working-age men available to meet the demand for the work they traditionally perform, no change in the gender division of labor will occur"* (p. 125, emphasis original). Thus, what we are looking for are structural demands that outrun the supply of male labor.

Chafetz lists four different kinds of processes that can unintentionally produce changes in the structure of gender inequality: population growth or decline, changes in the sex ratio of the population, and technological innovations and changes in the economic structure. The processes and their effects are listed below:

- *Population changes:* If the number of jobs that need to be filled remains constant, then the greater the growth in the working age population, the lower will be women's workforce participation. The inverse is true as well: As the size of the working population declines, women will gain greater access to traditionally male jobs, if the number of jobs remains constant.
- *Sex ratio changes:* The sex ratio of a population, the number of males relative to the number of females, tends to change under conditions of war and migration. If in the long run there is a reduction in the sex ratio (more women than men) of a population, women will gain access to higher paying and more prestigious work roles. Conversely, if there is an increase in the sex ratio, the restrictions on women's workforce participation will tend to increase.

- *Economic and technological changes:* There are two general features of men's and women's bodies that can influence women's workforce participation: Men on average tend to be stronger than women, and women carry, deliver, and nurse babies. When technological innovations alter strength, mobility, and length of employment requirements, then there are possibilities for gender change in those jobs. Women will tend to gain employment if new technologies reduce strength, mobility, and time requirements. In addition to work requirements, technology can also change the structure of the job market. As the economy expands due to technological innovations, women will tend to achieve greater workforce participation (holding population growth constant). On the other hand, if the economy contracts for whatever reason, women will tend to lose resource-generating work roles.

Intentional Change

In addition to unintentional change processes, Chafetz argues that there are specific ways in which people can act that help to address gender inequality. But before we get into those processes, we need to note that Chafetz maintains that gender change is particularly difficult for two reasons. First, women have more cross-cutting influences than any other group. Think about it this way: Almost every social group is gendered. This means that women are black, white, Chicano, Baptist, Buddhist, Jewish, pagan, homosexual, bisexual, heterosexual, homeless, professional, and so on. Women thus "differ extensively on all social variables except gender" (Chafetz, 1990, p. 170). These cross-cutting group affiliations make it extremely difficult to form a woman's ideology that doesn't cut across or offend some of the women the ideology is trying to embrace. The second issue is that women, unlike many disenfranchised groups, do not live in segregated neighborhoods. Chafetz argues that this reduces women's political clout because they are "dispersed throughout all electoral districts" (p. 171). Because of these difficulties, gender change may occur more slowly or diffusely than other types of change.

Chafetz's model of gender change is depicted in Figure 7.2. As we talk through Chafetz's theory, be sure to trace through the effects on the model. There are two major independent variables in Chafetz's theory: macro-structural changes and elite support. Notice that the primary issues pushing intentional change exist outside of the direct control of women. Look at Figure 7.2 and see where women's movements come into play. As you can see, the political activities of women are important, but they work more as a catalyst for change rather than the engine of it.

Chafetz's approach is similar to the way Karl Marx saw the production of class consciousness. According to Marxian theory, structure leads social change. Class consciousness is necessary for social change but it is produced as dialectical elements of the economic structure, such as increasingly disruptive business cycles and overproduction, play themselves out. For Marx, it is the *structure* that pushes people together and gives them the ability to see their oppression, communicate with one another, and mount the resistance that will lead to the demise of capitalism.

The macro-structural changes that Chafetz focuses on are industrialization, urbanization, and the size of the middle class. Chafetz claims that historically

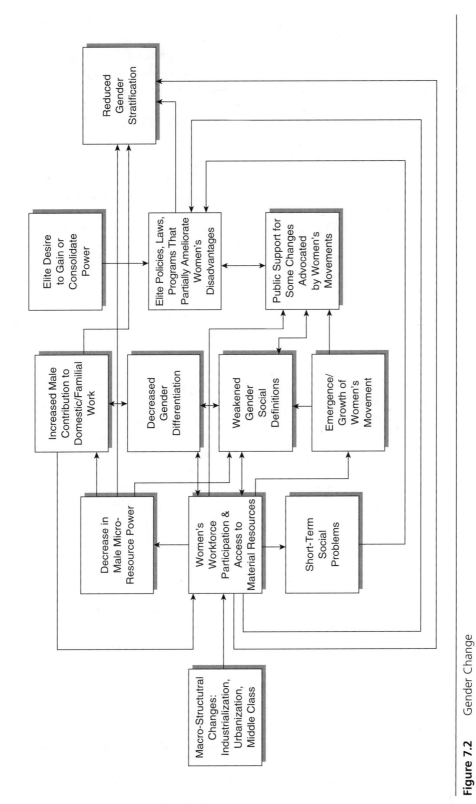

Figure 7.2   Gender Change

SOURCE: From Chafetz, J. S., *Gender Equity: An Integrated Theory of Stability and Change*, copyright © 1990, Sage Publications, Inc. Reprinted with permission.

almost all women's movements have been led by middle-class women. These women are among the first to experience gender consciousness due to the effects of industrialization and urbanization. Industrialization initiates a large number of social changes such as increases in urbanization, commodification, the use of money and markets, worker education, transportation and communication technologies, and so forth. These all work together to expand the size of the middle class. The importance of this expansion isn't simply that there are more middle-class people; it also means that there are more middle-class jobs available. Many of these are nondomestic jobs that may be filled by women.

Industrialization thus structurally creates workforce demands that women can fill. Women are more likely to be called upon to fill these roles when the number of males is kept fairly constant or at least the number doesn't increase at the same rate as the demand for labor. The more rapid the growth of industrialization and urbanization, the more likely the demand for labor will outpace the supply of men.

As women increase their workforce participation, they increase their level of material and political resources as well, thus decreasing males' relative micro power and the level of gender differentiation, as well as weakening gender stereotypes, ideologies, and normative expectations (gender social definitions). In addition, as women's resources and thus their micro power increase, they are more able to influence the household division of labor and men's contribution to familial and domestic work. And, as men contribute more to domestic and familial work, women are more able to gain resources through workforce participation, as noted by the feedback arrow in Figure 7.2.

In addition to influencing these micro-level issues, industrialization, urbanization, and women's increased workforce participation do two other things: They increase women's experience of relative deprivation and the number of women's social contacts. While absolute deprivation implies uncertain survival, relative deprivation is a subjective, comparative sense of being disadvantaged.

Social movements are extremely unlikely with groups that experience absolute deprivation. They have neither the will nor the resources to organize political movements. Relative deprivation, on the other hand, implies a group that has resources and experiences rising expectations. Thus, women who are newly moving into the workforce will tend to experience relative deprivation. They will begin to see that their salaries are not comparable to men's, and they will begin to accumulate resources that can potentially be used to become politically active.

Urbanization and industrialization also increase a group's ability to organize. Ralf Dahrendorf (1957/1959) talked about this ability to organize as the principal difference between quasi-groups and interest groups. Quasi-groups are those collectives that have latent identical role interests; they are people that hold the same structural position and thus have similar interests but do not experience a sense of "belongingness." Interest groups, on the other hand, "have a structure, a form of organization, a program or goal, and a personnel of members" (p. 180).

The interest group's identity and sense of belonging are produced when people have the ability to communicate, recruit members, form leadership, and create a unifying ideology. Urbanization and industrialization structurally increase the probability that these conditions of interest group membership will be met. Women living and working in technically advanced urban settings are more likely to come into contact with like others who are experiencing relative deprivation and the status and role dilemmas that come from women working a double workday.

As women begin to organize and as traditional gender definitions become weakened, public support for change is likely to arise. Chafetz argues that a significant portion of what women's movements have been able to achieve is related to articulated critical gender ideologies and radical feminist goals. While the public may not buy into a radical ideology or its set of goals, the ideologies and goals of women's movements will tend to justify gender change more broadly. This support, along with pressure from short-term social problems and a direct effect from women's workforce participation, place pressure upon elites to create laws, policies, and programs to help alleviate the unequal distribution of scarce resources by gender.

Let's pause a moment and talk about short-term social problems. Chafetz argues that as a result of women having and using greater levels of resources, short-term social problems are likely to arise. Chafetz uses the term "social problem" in a general sense to indicate the challenges that society at large have to overcome anytime significant change occurs. In other words, change to the social system brings a kind of disequilibrium that has to be solved so that actions and interactions can once again be patterned. Such social problems tend to accompany any type of social change as a society adjusts culturally and socially. In the case of gender, the short-term problems are related to women having greater levels of resources and thus higher levels of independence and power. Examples of these kinds of problems include increases in the divorce rate and women's demands for control over their own bodies. Social disruptions such as these tend to motivate elite support of women's rights in order to restore social order.

As you can see from Figure 7.2, elite support shows up twice in Chafetz's theory. Sociologists have learned that every social movement eventually requires support from the elite. The elite not only pass laws and oversee their enforcement, they can also lend other material or political support, such as money and social capital (networks of people in powerful positions).

The first place elites show up is in the way we have been talking about so far. Elites create new or support already established laws that help bring social order. In this case, the support comes mostly because elites perceive some form of social disorder. In other words, "they may perceive that basic problems faced by their society, which negatively affect large numbers of people and may possibly jeopardize their incumbency in elite roles, are exacerbated by a gender system that devalues and disadvantages women" (Chafetz, 1990, p. 152). This kind of change may be incremental and not specifically associated with women's movements.

The second place elite support shows up is as an exogenous variable (that is, one whose value is determined outside the model in which it is used), in the upper top right of the model. In this case, elite support may appear more directly tied to women's interests, though it may not necessarily be generated out of concern for women. Chafetz argues that we've made the mistake before of thinking of the elite as a single group with similar interests. In fact, the elite are divided into different factions, each struggling over power. In this kind of political environment, some elite groups are likely to see women as a potential resource and make promises designed to gather their support. At least some of these promises result in actual changes that reduce gender stratification.

I'm going to restate Chafetz's theory of gender change in brief propositions, since this is the discourse in which Chafetz works. Be sure to follow the elements of the propositions through the model. This way you'll be able to get a textual and visual rendering of the ideas.

- Taken together, the level of reduction in gender stratification due to intentional efforts is a positive function of the level of male domestic labor, elite support, and women's control of material resources; and a negative effect of the level of male micro power.
- Women's level of control over material resources is a positive function of industrialization, urbanization, and the size of the middle class. Elite support in general is a positive function of the level of women's control over material resources, the level of short-term social problems, the level of public support for changes advanced by women's movements, and elite competition.

Summary

- In general, Chafetz argues that workforce participation and control over material resources both stabilize and change a system of gender inequality. Gender inequality is perpetuated when women's participation in the workforce is restricted and reduced when women are allowed to work and control material resources.
- Chafetz argues that gender is stabilized more through voluntaristic actions rather than the use of coercive power. When men control the division of labor in society, they are able to exercise authority at the meso level through assuring male incumbency in elite positions and at the micro level through women's exchange of deference and compliance for material resources. Men thus control gender social definitions that set up engenderment processes: psychodynamic structuring, gender socialization, and the idealized expression of gender through impression management and interaction. Engenderment and wifely compliance work in turn to solidify women's exclusion from the workforce, to impose double duty upon those women who do work, to stabilize the unequal distribution of opportunities and resources, and to define negative worker attributes for women but positive ones for men.
- Gender inequality is reduced as women are allowed greater participation in the workforce and increased control over material resources. These factors decrease women's reliance upon men and men's authority over women. Men then contribute more to domestic and familial work, which further frees women to participate in the workforce and weakens gender stereotypes, norms, and ideologies. In addition, women's access to resource-generating work roles increases the probability of women's political movements, which

along with weakened gender definitions positively impacts public opinion and elite support for women. Short-term social issues that come about because of women's increased workforce participation also impact elite support, as well as the elite's desire to consolidate or gain political power. Gender stratification is reduced as women move into the workforce and control more material resources, as elites support women's rights and legislation, and as men's micro resource power is reduced and their domestic contribution increased.

Building Your Theory Toolbox

Learning More—Primary Sources

- Chafetz's most important work is *Gender equity: An integrated theory of stability and change,* Sage, 1990.
- She is also the editor of the excellent *Handbook of the sociology of gender,* Kluwer, 1999.

Check It Out

- *Web Bytes:* Janet Chafetz gives us a generalized, sociological explanation of gender oppression. The Web Byte for this chapter, *Randall Collins and Conflict Theory*, focuses on Chafetz's theoretical approach. In terms of different approaches to gender, there are Chapters 17 and 18, as well as the Web Byte *Patricia Hill Collins and Intersecting Oppressions*.
- In 1993, Chafetz joined Rae Lesser Blumberg, Scott Coltrane, Randall Collins, and Jonathan Turner to produce a synthesized theory of gender stratification. This is a rare and powerful effort: "Toward an integrated theory of gender stratification," *Sociological Perspectives, 36,* 185–216.

Seeing the World

- After reading and understanding this chapter, you should be able to answer the following questions (remember to answer them *theoretically*):
  o Explain why the workforce of women is so important to Chafetz's theory.
  o Explain how exchange processes work to produce gender inequalities. How do men use their micro power to their gendered advantage?
  o What are the three types of gender definitions?
  o Explain the differences between the intrapsychic structures of boys and girls. Explain how they are created and how they influence gender inequality.
  o How does social learning theory and dramaturgy contribute to our understanding of how gender inequality is voluntaristically reproduced?
  o How does gender inequality unintentionally change?
  o According to Chafetz, what two characteristics about gender make change difficult? (Be sure to explain them fully.)

o What are the structural forces that make gender change likely?
o How do women's movements form and how do they influence gender change? What does this tell us about social movements in general?

Engaging the World

- Consult at least four reliable Internet sources for the "separation of spheres." What is it and how does it figure into Chafetz's theory? List at least six ways that this historical, structural issue influences your life.
- Use your favorite search engine. Type in "global gender inequality." Based on information from at least three different societies, prepare a report on them. Also, compare and contrast these societies with the one you live in. How applicable do you think Chafetz's theory would be in those other three societies?
- Volunteer at a local women's shelter or resource center.

Weaving the Threads

- Use Mead's theory of the self and Collins's theory of interaction ritual to elaborate Kanter's ideas about organizational personality.
- Compare and contrast Collins's and Chafetz's theories of inequality. Think specifically about how Collins's theory might (or might not) contain issues of coercive and voluntaristic structures.
- Using Berger and Luckmann's theory, explain how inequality becomes reality for the women in Chafetz's theory. In other words, what are the cultural and constructivist components in Chafetz's theory? Can Berger and Luckmann add anything to our understanding of how inequality works?

CHAPTER 8

The Replication of Class

Pierre Bourdieu (1930–2002)

Photo: © Corbis.

Why are you in school? If you're like most students, you're in school to get a degree so you can get a job. Is the school that you're attending Harvard, Yale, or Princeton? Degrees from those schools are much more valuable in terms of getting a job. So, if you're in school so you can get a good job, why wouldn't you be at one of the Ivy League schools? I imagine that you've had enough sociology by now to tell me that if you're not at a top-10 school, it's probably because of class.

So, what is class and how is it reproduced? The first part of that question is easier to answer than the second (yet measuring class is difficult as well). We usually think of class in terms of socioeconomic-status (SES), as reflected by a measurement of wealth, income, occupational status, and education. And we usually think that class is replicated by the structuring of those factors.

But what if something more is going on? What if class is reproduced—structured—by something hidden, by something more insidious than SES? Further, what are structures anyway? How do they work to replicate class? And where do they come from in the first place? The work of Pierre Bourdieu will challenge what most of us think of as class. And he'll challenge the general sociological notion of structure.

Class is one of the most fundamental and important social categories—it determines not only how much we can spend at the mall, but also the life chances of every human being living under capitalism. Yet, according to Bourdieu, we have only just begun to understand how it works.

The Essential Bourdieu

Biography

Pierre Bourdieu was born August 1, 1930, in Denquin, France. Bourdieu studied philosophy under Louis Althusser at the École Normale Supérieure in Paris. After his studies, he taught for 3 years, 1955–1958, at Moulins. From 1958 to 1960, Bourdieu did empirical research in Algeria (*The Algerians,* 1962) that laid the groundwork for his sociology. In his career, he published over 25 books, one of which, *Distinction: A Social Critique of the Judgment of Taste,* was named one of the 20th century's 10 most important works of sociology (International Sociological Association). He was the founder and director of the Centre for European Sociology, and he held the French senior chair in sociology at Collège de France (the same chair held by sociologist and anthropologist Marcel Mauss). Craig Calhoun (2003) writes that Bourdieu was "the most influential and original French sociologist since Durkheim" (p. 274).

Passionate Curiosity

Bourdieu's passion is intellectual honesty and rigor. He of course is concerned with class, and particularly the way class is created and recreated in subtle, nonconscious ways. But above and beyond these empirical concerns is a driving intellect bent on refining critical thinking and never settling on an answer: "An invitation to think with Bourdieu is of necessity an invitation to think beyond Bourdieu, and against him whenever required" (Wacquant, 1992, p. xiv).

Keys to Knowing

constructivist structuralism, field, habitus, cultural capital, symbolic capital, linguistic markets, symbolic violence

---

Bourdieu's Perspective: Constructivist Structuralism

Sociology as Combat Sport

One of the wonderful things about working in academia is that it is part of my job to learn things. In preparing to write this chapter, I read Craig Calhoun's (2003) chapter on Bourdieu. The first section is entitled "Taking Games Seriously." In it, Professor Calhoun talks about Bourdieu's life as a former rugby player and how it influenced his theory. Throughout Bourdieu's writing, he uses such terms as field, game, and practice, and he talks about the bodily inculcation of culture. I had read Bourdieu and approached such terms and ideas as theoretical concepts. For some reason, it never occurred to me to understand their use as a kind of analogy— the analogy of the game. But as I read Calhoun's three pages about Bourdieu's fascination with rugby, his ideas and terms all came alive for me in a new way.

So, thanks to Craig Calhoun, the first thing we will talk about in Bourdieu's perspective is the analogy of the game. It's important to keep in mind that Bourdieu's use of the analogy doesn't come from a background in playing cards. Bourdieu was a rugby player. Rugby is a European game somewhat like American football, but it is considered by most to be much more grueling than football. In rugby, the play is continuous with no substitutions or time-outs (even for injury). The game can take anywhere from 60–90 minutes, with two halves separated by a 5-minute halftime. An important part of the game is the scrum. In a *scrum,* eight players from each side form a kind of inverted triangle by wrapping their arms around each other. The ball is placed in the middle and the two bound groups of men struggle head to head against each other until the ball is freed from the scrum. To see the struggle of the scrum gives a whole new perspective on Bourdieu's idea of social struggle.

Rugby matches take place on a field, involve strategic plays and intense struggles, and are played by individuals who have a clear physical sense of the game. Matches are of course structured by the rules of the game and the field. The field not only delineates the parameters of the play, but each field is different and thus knowledge of each field of play is important for success. The rules are there and, like in all games, come into play when they are broken, but a good player embodies the rules and the methods of the game. The best plays are those that come when the player is in the "zone," or playing without thinking. Trained musicians can also experience this zone by jamming with other musicians. Often when in such a state, the musician can play things that she or he normally would not be able to, and might have a difficult time explaining after the fact. The same is true for athletes. There is more to a good game than the rules and the field; the game is embodied in the performer. And, finally, there is the struggle, not only against the other team, but also the limitations of the field, rules, and one's own abilities.

You may not know it, but I just gave you a brief overview of Bourdieu's theory through the use of analogy. I will occasionally mention the game analogy as we work our way through the material, but I think if you keep it consistently in mind, you'll find it much easier to grasp the intent of Bourdieu's thinking.

Overcoming Dichotomies

There is a central dichotomy in sociology and the human sciences in general. This dichotomy, and those derived from it, sets some of the basic parameters of our discipline, such as the distinction between quantitative and qualitative methods and the divergence between structuralism and interactionism. This dichotomy also sets up one of the thorniest issues sociologists address: the link between the micro and macro levels of society. The dichotomy that I am referring to is the dilemma of structure (objective) versus agency (subjective), or, as Bourdieu talks about it, social physics versus social phenomenology. He characterizes the dichotomy between objective and subjective knowledge as one of the most harmful in the social sciences. And Bourdieu (1985) sees overcoming the break between objective and subjective knowledge as the most steadfast and important factor guiding his work (p. 15).

Bourdieu brings the two sides of the dichotomy together in what he calls **constructivist structuralism,** or structuralist constructivism—Bourdieu uses the term both ways—in which both structure and agency are given equal weight. Bourdieu (1989) says that within the social world, there are "objective structures independent of the consciousness and will of agents, which are capable of guiding and constraining their practices or their representations" (p. 14). He thus keeps structures in the objective social world that Durkheim gave us (although Bourdieu thinks a little differently about how and where these structures exist).

Yet, at the same time, Bourdieu emphasizes the constructivist and subjective sides. In Bourdieu's (1989) scheme, the subjective side is also structured in terms of "schemes of perception, thought, and action" (p. 14), which he calls *habitus*. Part of what Bourdieu does is to detail the ways through which both kinds of structures are constructed; thus, there is a kind of double structuring in Bourdieu's theory and research. But Bourdieu doesn't simply give us an historical account of how structures are produced. His theory also offers an explanation of how these two structures are dialectically related and how the individual uses them strategically in linguistic markets.

Dialecticism and a Theory of Practice

In preserving both sides of the dichotomy, Bourdieu has created a unique theoretical problem. He doesn't want to conflate the two sides as Giddens (Chapter 12) does, nor does he want to link them up in the way that Collins and Blau do. He wants to preserve the integrity of both domains and yet he characterizes the dichotomy as harmful. Bourdieu is thus left with a sticky problem: How can he keep and yet change the dichotomy between the objective and subjective moments without linking them together or blending them together?

Let's take this issue out of the realm of theory and state it in terms that are a bit more approachable. The problem that Bourdieu is left with is the relationship between the individual and society. Do we have free choice? Bourdieu would say yes. Does society determine what we do? Again Bourdieu would say yes. How can something be determined and yet the product of free choice? I've stated the issue a bit too simplistically for Bourdieu's theory, but I want you to see the problem clearly. Structure and agency, or the objective and subjective moments, create tension because they stand in opposition to one another. And that tension is exactly how Bourdieu solves his theoretical problem.

Bourdieu argues that the objective and constructive moments stand in a dialectical relationship (see Chapter 2 for the dialectic). Bourdieu's dialectic occurs between what he calls the field and the habitus. Both are structures; habitus is "incorporated history" and the field is "objectified history" (Bourdieu, 1980/1990, p. 66). We will go into more depth later on, but for now think of habitus as that part of society that lives in the individual as a result of socialization and think of the field as social structures. They are both much more, but what I want us to see now is Bourdieu's dialectic. The tension of the dialectic is between the subjective and objective structures and the dialectic itself is found in the individual and collective

necessarily build upon or include his earlier work. Bourdieu doesn't like "professorial definitions" but much prefers the idea of "open concepts" (Bourdieu & Wacquant, 1992, p. 95). Reflexive sociology, then, is more concerned with insight and inspiration than creed and rigor. Bourdieu gives us an inspiring way through which to bridge the gap between objective and subjective sociology. He gives us a dynamic space from which to think about the relations between structure and agency. And, as we will see more clearly in a moment, he gives us new ways to think about how class and social positions are patterned and how they change over time.

Concepts and Theory: Structuring Class

Capitals

The basic fact of capitalism is capital. Capital is different from either wealth or income. Income is generally measured by annual salary and wealth by the relationship between one's assets and debt. Both income and wealth are in a sense static; they are measurable facts about a person or group. Capital, on the other hand, is active: It's defined as accumulated goods devoted to the production of other goods. The entire purpose of capital is to produce more capital.

Bourdieu actually talks about four forms of capital—economic, social, symbolic, and cultural—all of which are invested and used in the production of class. Bourdieu uses *economic capital* in its usual sense. Economic capital is generally determined by one's wealth and income. As with Marx, Bourdieu sees economic capital as fundamental. However, unlike Marx, Bourdieu argues that the importance of economic capital is that it strongly influences an individual's level of the other capitals, which, in turn, have their own independent effects. In other words, economic capital starts the ball rolling; but once things are in motion, other issues may have stronger influences on the perpetuation of class inequalities.

Social capital refers to the kind of social network an individual is set within. It refers to the people you know and how they are situated in society. The idea of social capital can be captured in the saying, "it isn't what you know but who you know that counts." The distribution of social capital is clearly associated with class. For example, if you are a member of an elite class, you will attend elite schools such as Phillips Andover Academy, Yale, and Harvard. At those schools, you would be afforded the opportunity to make social connections with powerful people—for example, in elections over the past 30 years, there has been at least one Yale graduate running for the office of President of the United States. But economic capital doesn't exclusively determine social capital. We can build our social networks intentionally, or sometimes through happenstance. For example, if you attended Hot Springs High School in Arkansas during the early 1960s, you would have had a chance to become friends with Bill Clinton.

Symbolic capital is the capacity to use symbols to create or solidify physical and social realities. With this idea, Bourdieu begins to open our eyes to the symbolic nature of class divisions. Social groups don't exist simply because people decide to gather together. Max Weber recognized that there are technical conditions that must be met for a loose collection of people to form a social group: People must be able to communicate and meet with one another; there must be recognized leadership; and a group needs clearly articulated goals to organize. Yet, even meeting those conditions doesn't alone create a social group. Groups must be symbolically recognized as well.

With the idea of symbolic capital, Bourdieu pushes us past analyzing the use of symbols in interaction (Chapter 1). Symbolic interactionism argues that human beings are oriented toward meaning and meaning is the emergent result of ongoing symbolic interactions. We're symbolic creatures, but meaning doesn't reside within the symbol itself; it must be pragmatically negotiated in face-to-face situations. We've learned a great deal about how people create meaning in different situations because of symbolic interactionism's insights. But Bourdieu's use of symbolic capital is quite different.

Bourdieu recognizes that all human relationships are *created symbolically* and not all people have equal symbolic power. For example, I write a good number of letters of recommendation for students each year. Every form I fill out asks the same question: "Relationship to applicant?" And I always put "professor." Now, the *meaning* of the professor–student relationship emerges out of my interactions with my students, and my student–professor relationships are probably somewhat different from some of my colleagues as a result. However, neither my students nor I *created* the student–professor relationship.

Recall our earlier discussion about Bourdieu's critique of sociology's dichotomy. Here we can see both Bourdieu's critique and his answer: Social phenomenology can't account for the creation of the categories it uses, and social physics reifies the categories—Bourdieu (1991) tells us that objective categories and structures, such as class, race, and gender, are generated through the use of symbolic capital: "Symbolic power is a power of constructing reality" (p. 166).

Bourdieu (1989) characterizes the use of symbolic capital as both the power of constitution and the power of revelation—it is the power of "world-making . . . the power to make groups. . . . The power to impose and to inculcate a vision of divisions, that is, the power to make visible and explicit social divisions that are implicit, is political power par excellence" (p. 23). This power of world-making is based on two elements. First, there must be sufficient recognition to impose recognition. The group must be recognized and symbolically labeled by a person or group that is officially recognized as having the ability to symbolically impart identity, such as scientists, legislators, or sociologists in our society. Institutional accreditation, particularly in the form of an educational credential (school in this sense operates as a representative of the state), "frees its holder from the symbolic struggle of all against all by imposing the universally approved perspective" (p. 22).

The second element needed to world-make is some relation to a reality—"symbolic efficacy depends on the degree to which the vision proposed is founded in reality" (p. 23). I think it's best to see this as a variable. The more social or physical reality is already present, the greater will be the effectiveness of symbolic capital. This is the sense in which symbolic capital is the power to consecrate or reveal. Symbolic power is the power to reveal the substance of an already occupied social space. But note that granting a group symbolic life "brings into existence in an instituted, constituted form . . . what existed up until then only as . . . a collection of varied persons, a purely additive series of merely juxtaposed individuals" (p. 23). Thus, because legitimated existence is dependent upon symbolic capacity, an extremely important conflict in society is the struggle over symbols and classifications. The heated debate over race classification in the U.S. 2000 census is a good example.

There is a clear relationship between symbolic and cultural capital. The use of symbolic capital creates the symbolic field wherein cultural capital exists. In general, **cultural capital** refers to the informal social skills, habits, linguistic styles, and tastes that a person garners as a result of her or his economic resources. It is the different ways we talk, act, and make distinctions that are the result of our class. Bourdieu identifies three different kinds of cultural capital: objectified, institutionalized, and embodied. *Objectified cultural* capital refers to the material goods (such as books, computers, and paintings) that are associated with cultural capital. *Institutionalized cultural capital* alludes to the certifications (like degrees and diplomas) that give official acknowledgement to the possession of knowledge and abilities. *Embodied cultural capital* is the most important in Bourdieu's scheme. It is part of what makes up an individual's habitus, and it refers to the cultural capital that lives in and is expressed through the body. This function of cultural capital manifests itself as taste.

Taste refers to an individual preference or fondness for something, such as "he has developed a taste for expensive wine." What Bourdieu is telling us is that our tastes aren't really individual; they are strongly influenced by our social class—our tastes are embodied cultural capital. Here a particular taste is legitimated, exhibited, and recognized by only those who have the proper cultural code, which is class specific. To hear a piece of music and classify it as baroque rather than elevator music implies an entire world of understandings and classification. Thus, when individuals express a preference for something or classify an object in a particular way, they are simultaneously classifying themselves. Taste may appear as an innocent and natural phenomenon, but it is an insidious revealer of position. As Bourdieu (1979/1984) says, "Taste classifies, and it classifies the classifier" (p. 6). The issue of taste is "one of the most vital stakes in the struggles fought in the field of the dominant class and the field of cultural production" (p. 11).

Habitus

Taste is part of habitus and habitus is embodied cultural capital. Class isn't simply an economic classification (one that exists because of symbolic capital), nor is it merely a set of life circumstances of which people may become aware (class consciousness)—class is inscribed in our bodies. **Habitus** is the durable organization of one's body and its deployment in the world. It is found in our posture, and our way of walking, speaking, eating, and laughing; it is found in every way we use our body. Habitus is both a system whereby people organize their own behavior and a system through which people perceive and appreciate the behavior of others.

Pay close attention: This system of organization and appreciation is *felt* in our bodies. We physically feel how we should act; we physically sense what the actions of others mean, and we approve of or censure them physically (we are comfortable or uncomfortable); we physically respond to different foods (we can become voracious or disgusted); we physically respond to certain sexual prompts and not others—the list can go on almost indefinitely. Our humanity, including our class position, is not just found in our cognitions and mental capacity; it is in our very bodies.

One way to see what Bourdieu is talking about is to recall the rugby analogy. I love to play sand volleyball, and I only get to play it about once every 5 years, which means I'm not very good at it. I have to constantly think about where the ball and other players are situated. I have to watch to see if the player next to me is going for the ball or if I can do so. All this watching and mental activity means that my timing is way off. I typically dive for the ball 1.5 seconds too late, and I end up with a mouthful of sand rather than the ball (but the other bunglers on my team are usually impressed with my effort). Professional volleyball players compete in a different world. They rarely have to think. They sense the ball and their teammates and they make their moves faster than they could cognitively work through all the particulars. Volleyball is inscribed in their bodies.

Explicating what he calls the Dreyfus model, Bent Flyvbjerg (2001) gives us a detailed way of seeing what is going on here. The Dreyfus model indicates that there are five levels to learning: novice, advanced beginner, competent performer, proficient performer, and expert. Novices know the rules and the objective facts of a situation; advanced beginners still have concrete knowledge but see it contextually; and the competent performer employs hierarchical decision-making skills and feels responsible for outcomes. With proficient performers and experts, we enter another level of knowledge. The first three levels are all based on cognitions, but in the final two levels knowledge becomes embodied. Here situations and problems are understood "intuitively" and require skills that go beyond analytical rationality. With experts, "their skills have become so much a part of themselves that they are not more aware of them than they are of their own bodies" (p. 19).

Habitus thus works below the level of conscious thought and outside the control of the will. It is the embodied, nonconscious enactment of cultural capital that gives habitus its specific power, "beyond the reach of introspective scrutiny or control by the will . . . in the most automatic gestures or the apparently most insignificant techniques of the body . . . [it engages] the most fundamental principles of construction and evaluation of the social world, those which most directly express the division of labour . . . or the division of the work of domination" (Bourdieu, 1979/1984, p. 466). Bourdieu's point is that we are all, each one, experts in our class position. Our mannerisms, speech, tastes, and so on are written on our bodies beginning the day we are born.

There are two factors important in the production of habitus: education and distance from necessity. In *distance from necessity,* necessity speaks of sustenance, the things necessary for biological existence. Distance from the necessities of life enables the upper classes to experience a world that is free from urgency. In contrast, the poor must always worry about their daily existence. As humans move away from that essential existence, they are freed from that constant worry, and they are free to practice activities that constitute an end in themselves. For example, you probably have hobbies. Perhaps you like to paint, act, or play guitar as I do. There is a sense of intrinsic enjoyment that comes with those kinds of activities; they are ends in themselves. The poorer classes don't have that luxury. Daily life for them is a grind, a struggle just to make ends meet. This struggle for survival and the emotional toll it brings are paramount in their lives, leaving no time or resources for pursuing hobbies and "getting the most out of life."

We should think of distance from necessity as a continuum. You and I probably fall somewhere in the middle. We have to be somewhat concerned about our livelihood, but we also have time and energy to enjoy leisure activities. The elite are on the uppermost part of the continuum, and it shows in their every activity. For example, why do homeless people eat? They eat to survive. And they are hungry enough, they might eat anything, as long as it isn't poisonous. Why do working classes or nearly poor people eat? For the same basic reason: The working classes are much better off than the homeless, but they still by and large live hand to mouth. However, because they are further removed from necessity, they can be more particular about what they eat, though the focus will still be on the basics of life, a "meat and potatoes" menu. Now, why do the elite eat? You could say they eat to survive, but they are never aware of that motivation. Food doesn't translate into the basics of survival. Eating for the elite classes is an aesthetic experience. For them, plate presentation is more important than getting enough calories.

Thus, the further removed we are from necessity, the more we can be concerned with abstract rather than essential issues. This ability to conceive of form rather than function—aesthetics—is dependent upon "a generalized capacity to neutralize ordinary urgencies and to bracket off practical ends, a durable inclination and aptitude for practice without a practical function" (Bourdieu, 1979/1984, p. 54). And this aesthetic works itself out in every area. In art, for example, the upper-class aesthetic of luxury, or what Bourdieu calls the *pure gaze*, prefers art that is abstract while the popular taste wants art to represent reality. In addition, distance from economic necessity implies that all natural and physical desires and responses are to be sublimated and dematerialized. The working class, because it is immersed in physical reality and economic necessity, interact in more physical ways through touching, yelling, embracing, and so forth than do the distanced elite. A lifetime of exposure to worlds so constructed confers cultural pedigrees, manners of applying aesthetic competences that differ by class position.

This embodied tendency to see the world in abstract or concrete terms is reinforced and elaborated through *education*. One obvious difference between the education of the elite and the working classes is the kind of social position in which education places us. The education system channels individuals toward prestigious or devalued positions. In doing so, education manipulates subjective aspirations (self-image) and demands (self-esteem). Another essential difference in educational experience has to do with the amount of rudimentary scholastics required—the simple knowing and recognizing of facts versus more sophisticated knowledge. This factor varies by number of years of education, which in turn varies by class position. At the lower levels, the simple recitation of facts is required. At the higher levels of education, emphasis is placed on critical and creative thought. At the highest levels of education, even the *idea* of "fact" is understood critically and held in doubt.

Education also influences the kind of language we use to think and through which we see the world. We can conceive of language as varying from complex to simple. More complex language forms have more extensive and intricate syntactical elements. Language is made up of more than words; it also has structure. Think about the sentences that you read in a romance novel and then compare them to those in an advanced textbook. In the textbook, they are longer and more complex, and that complexity increases as you move into more scholarly books. These more complex syntactical elements allow us to construct sentences that correspond to multileveled thinking—this is true because both writing and thinking are functions of language. The more formal education we receive, the more complex are the words and syntactical elements of our language. Because we don't just think *with* language, we think *in* language, the complexity of our language affects the complexity of our thinking. And our thinking influences the way in which we see the world.

Here's a simple example: Let's say you go to the zoo, first with my dog and then with three different people. You'd have to blindfold and muzzle my dog, but if you could get her to one of the cages and then remove the blinders, she would start barking hysterically. She would be responding to the content of the beasts in front of her. All she would know is that those things in front of her smell funny, look dangerous, and are undoubtedly capable of killing her, but she's going to go down fighting. Now picture yourself going with three different people, each from a different social class and thus education level. The first person has a high school education. As you stand in front of the same cage that you showed to my dog, he says, "Man, look at all those apes." The second person you go with has had some college education. She stands in front of the cage and says, "Gorillas are so amazing." The third person has a master's level education and says, "Wow, I've never seen gorilla gorilla, gorilla graueri, and gorilla berengei all in the same cage."

Part of our class habitus, then, is determined by education and its relationship to language. Individuals with a complex language system will tend to see objects in terms of multiple levels of meaning and to classify them abstractly. This type of linguistic system brings sensitivity to the structure of an object; it is the learned ability to respond to an object in terms of its matrix of relationships. Conversely, the less complex an individual's classification system, the more likely are the organizing syntactical elements to be of limited range. The simple classification system is characterized by a low order of abstractedness and creates more sensitivity to the *content* of an object, rather than its structure.

Bourdieu uses the idea of habitus to talk about the replication of class. Class, as I mentioned earlier, isn't simply a part of the social structure; it is part of our body. We are not only categorized as middle class (or working class or elite), we *act* middle class. Differing experiences in distance from necessity and education determine one's tastes, ways of seeing and experiencing the world, and "the most automatic gestures or apparently most insignificant techniques of the body—ways of walking or blowing one's nose, ways of eating or talking" (Bourdieu, 1979/1984, p. 466). We don't choose to act or not act according to class; it's the result of lifelong socialization. And, as we act in accordance with our class, we replicate our class. Thus, Bourdieu's notion of how class is replicated is much more fundamental and insidious than Marx's and more complex than Weber's.

However, we would fall short of the mark if we simply saw habitus as a structuring agent. Bourdieu intentionally uses the concept (the idea originated with Aristotle) in order to talk about the creative, active, and inventive powers of the agent. He uses the concept to get out of the structuralist paradigm without falling back into issues of consciousness and unconsciousness. In habitus, class is structured but it isn't completely objective—it doesn't merely exist outside of the individual because it's a significant part of her or his subjective experience. In habitus, class is *structured but not structuring*—because as with high-caliber athletes and experts, habitus is intuitive. The idea of habitus, then, shows us how class is replicated subjectively and in daily life, and it introduces the potential for inspired behaviors above and beyond one's class position. Indeed, the potential for exceeding one's class is much more powerful with Bourdieu's habitus than with conscious decisions—most athletes, musicians, and other experts will tell you that their highest achievements come under the inspiration of visceral intuition rather than rational processes. It is through habitus that the practices of the dialectic are performed.

Fields

As we talk about Bourdieu's notion of the *field*, keep the rugby analogy in mind. Just like in rugby, fields are delineated spaces wherein "the game" is played. Obviously, in Bourdieu's theoretical use of field, the parameters are not laid out using fences or lines on the ground. The parameters of the theoretical field are delineated by networks or sets of connections among objective positions. The positions within a field may be filled by individuals, groups, or organizations. However, Bourdieu is adamant that we focus on the relationships among the actors and not the agents themselves. It's not the people, groups, or even interactions that are important; it's the relationships among and between the positions that sets the parameters of a field. For example, while the different culture groups (like theater groups, reading clubs, and choirs) within a region may have a lot in common, they probably do not form a field because there are no explicit objective relationships among them. On the other hand, most all the universities in the United States do form a field. They are objectively linked through accreditation, professional associations, federal guidelines, and so forth. These relationships are sites of active practices; thus, the parameters of a field are always at stake within the field itself. In other words, because fields are defined mostly through relationships and relationships are active, which positions and relationships go into making up the field is constantly changing. Therefore, what constitutes a field is always an empirical question.

Fields are directly related to capitals. The people, groups, and organizations that fill the different objective positions are hierarchically distributed in the field, initially through the overall volume of all the capitals they possess and secondly by the relative weight of the two particular kinds of capital, symbolic and cultural. More than that, each field is different because the various cultures can have dissimilar weights. For example, cultural capital is much more important in academic rather than economic fields; conversely, economic capital is more important in economic fields than in academic ones. All four capitals or powers are present in each, but they aren't all given the same weight. It is the different weightings of

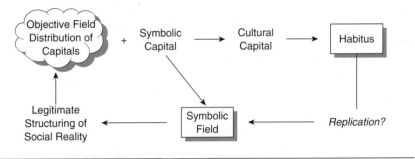

Figure 8.1   Habitus and the Replication of Class

the cultures that define the field, and it is the field that gives validity and function to the capitals.

While the parameters of any field cannot be determined prior to empirical investigation, the important consideration for Bourdieu is the correspondence between the empirical field and its symbolic representation. The objective field corresponds to a symbolic field, which is given legitimation and reality by those with symbolic capital. Here symbolic capital works to both construct and recognize—it creates and legitimates the relations between and among positions within the field. In this sense, the empirical and symbolic fields are both constitutive of class and of social affairs in general. It is the symbolic field that people use to view, understand, and reproduce the objective.

I've pictured Bourdieu's basic theory of class structuring in Figure 8.1. When reading the model, keep in mind Bourdieu's intent with open concepts. This model is simply a heuristic device—something we can use to help us see the world. The figure starts on the far left with the objective field and the distribution of capitals. But for Bourdieu, the objective field isn't enough to account for class reality and replication, and that is where many sociologists stop. Bourdieu takes it further in that the objective field becomes real and potentially replicable through the use of symbolic capital. The use of symbolic capital creates the symbolic field, which in turn orders and makes real the objective field. The exercise of symbolic capital, along with the initial distribution of capitals, creates cultural capital that varies by distance from necessity and by education. And cultural capital produces the internal structuring of class: habitus. But notice that the model indicates that the potential of habitus to replicate is held in question—it is habitus exercised in linguistic markets and symbolic struggles that decides the question.

Concepts and Theory: Replicating Class

Linguistic Markets

Bourdieu (1991) says that "every speech act and, more generally, every action" is an encounter between two independent forces (p. 37). One of those forces is habitus, particularly in our tendency to speak and say things that reveal our level of cultural capital. The other force comes from the structures of the linguistic market. A **linguistic market** is "a system of relations of force which impose themselves as a system of specific sanctions and specific censorship, and thereby help fashion linguistic production by determining the 'price' of linguistic products" (Bourdieu & Wacquant, 1992, p. 145).

The linguistic market is like any other market: It's a place of exchange and a place to seek profit. Here exchange and profit are sought through linguistic elements such as symbols and discourses. The notion of a free market is like an ideal type: It's an idea against which empirical instances can be measured. All markets are structured to one degree or another, and linguistic markets have a fairly high degree of structuring. One of the principal ways they are structured is through formal language.

Every society has formalized its language. Even in the case where the nation might be bilingual, such as Canada, the languages are still formalized. Standard language comes as a result of the unification of the state, economy, and culture. The education system is used to impose restrictions on popular modes of speech and to propagate the standard language. We all remember times in grammar school when teachers would correct our speech ("there is no such word as *ain't*"). In the university, this still happens, but mostly through the application of stringent criteria for writing.

Linguistic markets are also structured through various configurations of the capitals and the empirical field. As we've seen, empirical fields are defined by the relative weights of the capitals—so, for example, religious fields give more weight to symbolic capital and artistic fields more import to cultural capital, but they both need and use economic capital. The same is true with linguistic markets. Linguistic markets are defined through the relative weights of the capitals and by the different discourses that are valued. For example, the linguistic market of sociology is heavily based on cultural capital. And in order to do well in that market, you would have to know a fair amount about Karl Marx, Émile Durkheim, Michel Foucault, Pierre Bourdieu, Dorothy Smith, and so forth. Linguistic markets are also structured by the empirical field, in particular by the gaps and asymmetries that exist between positions in the field (by their placement and position of capitals, some positions in a field are more powerful than others). These empirical inequalities help structure the exchanges that take place within a linguistic market.

When people interact with one another, they perform speech acts—meaningful kinds of behaviors that are related to language. In a speech act, habitus and linguistic markets come together. In other words, the person's embodied class position and cultural capital are given a certain standing or evaluation within the linguistic market. The linguistic market contains the requirements of formal language; the salient contour of capitals; and the objective, unequal distribution of power within the empirical field.

Let me give you three examples from my own life. When I go to a professional conference, I present papers to and meet with other academics. My habitus has a number of different sources, among them are training in etiquette by a British mother and many years spent studying scholarly texts and engaging in academic discourse. The linguistic market in academia is formed by the emphasis on cultural and symbolic capital, and by the positions in the empirical field held by everyone at the conference; some people have more powerful positions and others less so. Each encounter, each speech act, is informed by these issues. In such circumstances, I tend to "feel at home" (habitus), and I interact freely, bantering and arguing with other academics in a kind of "one-upmanship" tournament.

This weekend, I will be going to the annual Christmas party at my wife's work. Here the linguistic market is different. Economic capital and the cultural capital that goes along with it are much more highly prized. And the empirical field is made up of differing positions and relationships achieved in the struggle of American business. Because of these differences, my market position is quite different here from what it was at the professional conference. In fact, I have no market position. Worse, my habitus remains the same. The way I talk—the words I use and the way I phrase my sentences—is very different from the other people at this event. The tempo of my speech is different (it's much slower) as is the way I walk and hold myself. In this kind of situation, I try and avoid speech acts. When encounters are unavoidable, I say as little as possible because I know that what I have to say, the way I say it, and even the tempo of my speech won't fit in.

These two different examples illustrate an extremely important point in Bourdieu's theory: Individuals in a given market recognize their institutional position, have a sense of how their habitus relates to the present market, and anticipate differing profits of distinction. In my professional conference example, I anticipate high rewards and distinction; but in the office party example, I anticipate low distinction and few rewards. In situations like the office example, anticipation acts as a *self-sanctioning mechanism* through which individuals participate in their own domination. Perhaps "domination" sounds silly with reference to an office party, but it isn't silly when it comes to job interviews, promotions, legal confrontations, encounters with government officials, and so forth. I gave you an example that we can relate to so that we could more clearly understand what happens in other, more important speech acts.

These kinds of speech acts are the arena of symbolic violence. In society, power is seldom used as coercive force, but is translated into symbolic form and thereby endowed with a type of legitimacy. Symbolic power is an invisible power and is generally misrecognized: We don't see it as power; we see it as legitimate. In recognizing as legitimate the hierarchical relations of power in which they are embedded, the oppressed are participating in their own domination: "All symbolic domination presupposes, on the part of those who submit to it, a form of complicity which is neither passive submission to external constraint nor a free adherence to values. . . . It is inscribed, in a practical state, in dispositions which are impalpably inculcated, through a long and slow process of acquisition, by the sanctions of the linguistic market" (Bourdieu, 1991, pp. 50–51).

My third example is from a conversation with my wife. In most conversations among equals, formal linguistic markets have little if any power. We talk and joke around, paying no attention to the demands of proper speech. I'm certain that you can think of multitudes of such speech acts: talking with friends at a café or at the gym or in your apartment. And those kinds of speech acts will always stay that way, unless one of you has a higher education or a greater distance from operate. Second, individuals and groups can challenge the subjective meanings intrinsic within the symbolic field. In daily speech acts, the individual can disrupt the normality of the symbolic field through insults, jokes, questions, rumors, and so on. Groups can also challenge "the way things are" by redefining history.

Building Your Theory Toolbox

Learning More—Primary Sources

- I recommend that you begin your exploration of Bourdieu with the following:
  - *Distinction: A social critique of the judgment of taste*, Harvard University Press, 1984.
  - Social space and symbolic power. *Sociological Theory, 7*, 14–25, 1989.
  - *Language and symbolic power*, Harvard University Press, 1991.
  - *An invitation to reflexive sociology*, University of Chicago Press, 1992.
  - *Acts of resistance: Against the tyranny of the market*, New Press, 1999.

Learning More—Secondary Sources

- There are a number of good secondary sources. The two I find most helpful are
  - David Swartz: *Culture and power: The sociology of Pierre Bourdieu*, University of Chicago Press, 1998.
  - Richard Jenkins: *Pierre Bourdieu* (Key Sociologists), Routledge, 2002.

Check It Out

- *Web Byte—Erik Olin Wright: Measuring Class Inequality*
- *Symbolic differences and identity:* Michele Lamont & Marcel Fournier (Eds.): *Cultivating differences: Symbolic boundaries and the making of inequality,* University of Chicago Press, 1993.
- *Overcoming dichotomies:* See Anthony Giddens in this book, Chapter 12.

Seeing the World

- After reading and understanding this chapter, you should be able to answer the following questions (remember to answer them *theoretically*):
  - Explain Bourdieu's constructivist structuralism approach. In your answer, be certain to define and explain the problems associated with the social physics (structural) and social phenomenology (subjective) approaches and how Bourdieu's approach counters both.
  - How are symbolic fields produced?
  - What is habitus and how is it produced?
  - How are class inequalities replicated, and how is class contingent? In your answer, be certain to explain linguistic markets, symbolic violence, and the role that habitus plays.

Engaging the World

- Use Bourdieu's theory to describe and explain the differences between the way you talk with your best friend versus the way you talk with your theory professor.
- Using Bourdieu's take on misrecognition, analyze the following ideas: race, gender, and sexuality. What kinds of things must we misrecognize in order for these to work as part of the symbolic violence of this society? Look up Bourdieu's notion of *doxa* (1972/1993, p. 3; 1980/1990, p. 68). How do doxa and symbolic violence work together in the oppression of race, gender, and sexuality?
- Choose the structure of inequality that you know best (race, gender, sexuality, religion). Using what you already know, analyze that structure using Figure 8.1. How would approaching the study of inequality change using Bourdieu?

Weaving the Threads

- Compare and contrast Bourdieu's constructivist structuralism with either Blau's or Collins's theory of the micro–macro link. Obviously, you will want to point out the individual similarities and differences, but also pay attention to the approaches themselves. What are the differences between approaching the structure/subject debate from a micro–macro link and thinking about it in terms of doing away with the dichotomy?
- In what ways are linguistic exchanges different from social exchanges (Blau)? How are they the same? What do we gain or lose by thinking of these as two different markets?
- Both Berger and Luckmann and Bourdieu give us theories based on culture. Pick at least two of the most important differences between their theories and explain them. Can these differences be brought together to form a fuller theory of culture? If so, how?
- Compare and contrast Bourdieu's ideas of linguistic markets and symbolic violence with Collins's (Chapter 5) idea of deference and demeanor rituals. How could these ideas be brought together to form a more robust theory of stratification in everyday encounters?

Wallerstein's Perspective: World-Systems Critique

Wallerstein uses a Marxian perspective to critique global capitalism. As such, there are two main strands in his perspective: Marxian theory and world systems. I'm going to hold off talking about his Marxian roots until we get to Wallerstein's theory. I think this will better help us see how his theory works. The idea of world systems, however, is a unique perspective and thus requires some explanation up front. We want to specifically address two issues: globality and historicity.

Globality

Wallerstein intentionally uses the hyphen in world-system to emphasize that he is talking about systems that constitute a world or a distinct way of existing. We touched on the idea of systems in the introduction to this section, but let me remind you that this approach looks at society as an interrelated whole, with every part systemically influencing the others. A systems approach additionally places emphasis on the relationship between the system and its environment. Thus, world-systems analysis argues that nations or collectives change in response to systemic factors that press upon it from the outside. For example, according to world-systems analysis, Peru isn't "modernizing" because it is something every nation will do; Peru is modernizing because it is caught in a global capitalist system that is pressuring it to change. Wallerstein captures this systemic approach with the idea of *globality*. With this idea, Wallerstein is introducing a distinct level of analysis. It is an analysis that looks at the relationships among nations and other political entities that form a system. Thus, the systemic factors cut across cultural and political boundaries and create an "integrated zone of activity and institutions which obey certain systemic rules" (Wallerstein, 2004, p. 17).

The idea of systems is important because it points out the differences between globality and globalization. *Globalization* is a term that was coined in the early 1990s to describe "the closer integration of the countries and peoples of the world which has been brought about by the enormous reduction of costs of transportation and communication, and the breaking down of the ratification barriers to the flows of goods, services, capital, knowledge, and (to a lesser extent) people across borders" (Stiglitz, 2003, p. 9). While the boundaries of what is to be included in the definition of globalization are unclear, it is clear that it is seen as an economic phenomenon, one that is focused on free trade. In this economy, as in most economists' models, market forces and invisible hands operate like devices of natural selection to control prices and the behaviors of firms. Wallerstein sees these forces as mystical and reified. Globality, on the other hand, explicates the systemic properties of both the economy and relations among nations.

Historicity

If social actors such as nations, institutions, and groups are related to each other through a specific system, then the history of that system is extremely important for understanding how the system is working presently. Wallerstein picked up the notions of structural time and cyclical process from French historian and educator Fernand Braudel. Braudel criticized event-dominated history, the kind with which we are most familiar, as being too idiographic and political.

The prefix "idio" specifically refers to the individual or one's own. Idiographic knowledge, then, is focused on unique individuals and their events. An example of this event history approach is to understand U.S. history in terms of things like Abraham Lincoln and the Civil War and Martin Luther King Jr. and the civil rights movement. Such an understanding doesn't see changes through history as the result of systematic social facts, but, rather, it perceives historical change as occurring through unique events and political figures.

Braudel felt that this kind of history is dust and tells us nothing about the true historical processes. Yet Braudel also criticized the opposite approach, nomothetic knowledge. The word *nomothetic* is related to the Greek word *nomos,* which means law. The goal in seeking nomothetic knowledge, like that of science, is to discover the abstract and universal laws that underpin the physical universe. According to Braudel, when nomothetic knowledge is sought in the social sciences, it more often than not creates mythical, grand stories that legitimate the search for universal laws instead of explaining historical social history.

Wallerstein's idea of *historicity* lies between the ideographic focus on events and the law-like knowledge of science. Rather than focusing on events, Wallerstein's approach concentrates on the history of structures within a world-system. For example, capitalism is a world-system that has its own particular history. There have always been people who have produced products to make a profit, but the capitalism of modernity, the kind that Weber (1904–1905/2002) termed "rational capitalism," is unique to a particular time period. An account of rational capitalism from its beginnings, from around the 16th century, that would include all the principal players (such as nations, firms, households, and so forth) and their systemic relations is what Wallerstein has in mind.

Historicity thus includes the unique variable of time. In taking account of world-systems rather than event history, historicity is centered upon *structural time* and the *cyclical time* within the structures. Wallerstein is telling us that structures have histories, and it is the history of structures with which we should be concerned, rather than events, because structures set the frames within which human behavior and meaning take place. Structures have life spans, they are born and they die, and within that span there are cyclical processes. Here we begin to see Wallerstein's Marxian roots clearly. The idea of structural change occurring through cycles comes from Marx's notion of the dialectic (please see Chapter 2 for an introduction to this important concept). As we move through Wallerstein's theory, keep in mind the idea of structural change through dialectical oppositions—pay attention to the contradictions that are intrinsic to capitalism.

Concepts and Theory: The Dialectics of Capitalism

Wallerstein's critique is essentially a Marxian critique. Marx did what Wallerstein says needs to be done: He focused on structures moving through cyclical time.

He was particularly interested in capitalism—and, according to Wallerstein, the elements of capitalism are in fact the only features that can truly create a world-system today. Certain of Marx's concepts, then, have special importance in explaining and critiquing the world-system. Among them are the division of labor, exploitation, accumulation, and overproduction. I will be explaining these concepts before putting them into place in Wallerstein's theory.

The Division of Labor and Exploitation

The division of labor is one of the more important elements in the Marxian perspective; in some ways, it is the most important. Marx starts his theory with the idea of species being. *Species being* contains two ideas: First, the way the human species exists is through creative production; second, humans become conscious of their existence (or being) through the mirror effect of the product. Humanity, then, is defined and knows itself through creative production. There is, then, an intimate connection between producer and product: *The very existence of the product defines the nature of the producer.* If you think about it, we acknowledge this connection every time we meet someone new. One of the first questions we ask a new person is, "What do you do?" In doing so, we assume the connection between what people do and who they are.

For Wallerstein, the importance of the *division of labor* is that it is the defining characteristic of an economic world-system. Labor, of course, is an essential form of human behavior; without it we would cease to exist. By extension, the division of labor creates some of the most basic kinds of social relationships; these relationships are, by definition, relations of dependency. In our division of labor, we depend upon each other to perform the work that we do not. I depend upon the farmer for food production, and the farmer depends upon teachers to educate her or his children. These relations of dependency connect different people and other social units into a structured whole or system. Wallerstein argues that the world-system is connected by the current capitalist division of labor: World-systems are defined "quite simply as a unit with a single division of labor and multiple cultural systems" (Wallerstein, 2000, p. 75). Multiple cultural systems are included because world-systems connect different societies and cultures.

The important feature of this division of labor is that it is based on exploitation. I went into some detail defining exploitation in Chapter 7, so I won't do so now. But as we go through the next section, keep in mind two things: First, exploitation is a measurable entity; it is the difference between what a worker gets paid and what she or he produces. Different societies can have different levels of exploitation. For example, if we compare the situation of automobile workers in the United States with those in Mexico, we will see that the level of exploitation is higher in Mexico. The second thing to keep in mind is that exploitation is fundamental to capitalism. Surplus labor and exploitation are the places from which profit comes and are thus necessary for capitalism.

What is important to see here is that profit is based on exploitation and there are limitations to exploitation. Yet the drive for exploitation doesn't let up; capitalists by definition are driven to increase profits. The search for new means of exploitation, then, eventually transcends national boundaries: *Capitalists export exploitation.* Because of the limitations on the exploitation of workers in advanced capitalist countries—due primarily to the effects of worker movements, state legislation, and the natural limitations of technological innovation—firms seek other labor markets where the level of exploitation is higher. Marx had a vague notion of this, but Wallerstein's theory is based upon it. It is the exportation of exploitation that structures the division of labor upon which the world-economy is based.

Accumulation and Overproduction

We all know what modern capitalism is: It is the investment of money in order to make more money (profit). As Wallerstein (2004) says, "we are in a capitalist system only when the system gives priority to the *endless* accumulation of capital" (p. 24, emphasis original). We see the drive to make money in order to make more money all around us; but most people only think about the personal effects this kind of capitalism has (like the fact that Bill Gates is worth $46.6 billion). But what are the effects on the economy? Most Americans would probably say that the effect on the economy is a good one: continually expanding profits and higher standards of living. Perhaps, but Wallerstein wants us to see that something else is going on as well. In order to fully understand what he has in mind, we need to think about the role of government in the endless pursuit of the accumulation of capital.

It's obvious that for capitalism to work, it needs a strong state system. The state provides the centralized production and control of money; creates and enforces laws that grant private property rights; supplies the regulation of markets, national borders, inter-organizational relations; and so forth. But there is something else that the state does in a capitalist system. We generally assume that the firm that pays the cost enjoys the benefits, as in the capitalist invests the money so she or he can enjoy the profit. However, the state actually decides what proportion of the costs of production will be paid by the firm. In this sense, capitalists are subsidized by the state.

There are three kinds of costs that the state subsidizes: the costs associated with transportation, toxicity, and the exhaustion of raw materials. Firms rarely if ever pay the full cost of transporting their goods; the bulk of the cost for this infrastructure is borne by the state, for such things as road systems. Almost all production produces toxicity, whether noxious gases, waste, or some kind of change to the environment. How and when these costs are incurred and who pays for them is always an issue. The least expensive methods are short-term and evasive (dumping the waste, pretending there isn't a problem), but the costs are eventually paid and usually by the state. Capitalist production also uses up raw materials; but again firms rarely pay these costs. When resources are depleted, the state steps in to restore or recreate the materials. Economists refer to the expenses of capitalist production that are paid by the state as *externalized costs,* and we will see that in this matter not all states are created equal.

However helpful these externalized costs are to the pursuit of accumulation, states that contain the most successful capitalist enterprises do more: They provide a structure for *quasi-monopolies.* A monopoly is defined as the exclusive control

a cycle that eventually leads to the demise of the empire. Maintaining a standing army that is geographically extended costs quite a bit of money. This money is raised through tribute and taxation. Heavy taxes make the system less efficient, in terms of economic production, and this increases the resistance of the populace as well. Increasing resistance means that the military presence must be increased, which, in turn, increases the cost, taxation, and resistance, and it further lowers economic efficiency. These cycles continue to worsen through structural time (see historicity above) until the empire falls. Examples of such world-empires include Rome, China, and India.

These world-empire cycles continued until about 1450, when a world-economy began to develop. Rather than a common political system, *world-economies* are defined through a common division of labor and through the endless accumulation of capital. As we saw earlier, in the absence of a political structure or common culture, the world-system is created through the structures intrinsic to capitalism. The worldwide division of labor created through the movement of products and labor from advanced capitalist nations to rising capitalist nations creates relationships of economic dependency and exploitation. These capitalist relationships are expressed through three basic types of economic states: core, semi-periphery, and periphery.

Briefly, *core states* are those that export exploitation; enjoy relatively light taxation; have a free, well-paid labor force; and constitute a large consumer market. The state systems within core states are the most powerful and are thus able to provide the strongest protection (such as trade restrictions) and capitalist inducements, such as externalizing costs, patent protection, tax incentives, and so on. *Periphery states* are those whose labor is forced (very little occupational choice and few worker protections) and underpaid. In terms of a capitalist economy and the world-system, these states are also the weakest—they are able to provide little in the way of tax and cost incentives, and they are the weakest players in the world-system. The periphery states are those to which capitalists in core states shift worker exploitation and more competitive, less profitable products. These shifts result in "a constant flow of surplus-value from the producers of peripheral products to the producers of core-like products" (Wallerstein, 2004, p. 28).

The relationship, then, between the core and the periphery is one of production processes and profitability. There is a continual shift of products and exploitation from core to periphery countries. Furthermore, there are cycles in both directions: Periphery countries are continually developing their own capitalist-state base. As we've seen, profitability is highest in quasi-monopolies and these, in turn, are dependent upon powerful states. Thus, changing positions in the capitalist world-economy is dependent upon the power of the state.

Overtime, periphery economies become more robust and periphery states more powerful: Worker protection laws are passed, wages increase, and product innovation begins to occur; the states can then begin to perform much like the states in core countries—they create tax incentives and externalize costs for firms, grant product protection, and they become a more powerful player in the world-system economy. These nations move into the semi-periphery. *Semi-periphery* states are those that are in transition from being a land of exploitation to being a core player, and they both export exploitation and continue to exploit within their own country.

A good illustration of this process is the textile industry. In the 1800s, textiles were produced in very few countries and it was one of the most important core industries; by the beginning of the 21st century, textiles had all but moved out of the core nations. A clear and recent example of this process is Nike. Nike is the world's largest manufacturer of athletic shoes, with about $10 billion in annual revenue. In 1976, Nike began moving its manufacturing concerns from the United States to Korea and Taiwan, which at the time were considered periphery states. Within 4 years, 90% of Nike's production was located in Korea and Taiwan.

However, both Korea and Taiwan were on the cusp, and within a relatively short period of time they had moved into the semi-periphery. Other periphery states had opened up, most notably Bangladesh, China, Indonesia, and Vietnam. So, beginning in the early 1990s, Nike began moving its operations once again. Currently, Indonesia contains Nike's largest production centers, with 17 factories and 90,000 employees. But that status could change. Just a few years ago, in 1997, the Indonesian government announced a change in the minimum wage, from $2.26 per day to $2.47 per day. Nike refused to pay the increase and in response, 10,000 workers went on strike. In answer to the strike, a company spokesperson, Jim Small, said, "Indonesia could be reaching a point where it is pricing itself out of the market" (Global Exchange, 1998).

Yet the existence of the semi-periphery doesn't simply serve as a conversion point; it has a structural role in the world-system. Because the core, periphery, and semi-periphery share similar economic, political, and ideological interests, the semi-periphery acts as a buffer that lessens tension and conflict between the core and periphery nations. "The existence of the third category means precisely that the upper stratum is not faced with the *unified* opposition of all the others because the *middle* stratum is both exploited and exploiter" (Wallerstein, 2000, p. 91, emphasis original).

Kondratieff Waves

Since 1450, world-economies have moved through four distinct phases. These phases occur in what are called Kondratieff waves (K-waves), named after Nikolai Kondratieff, a Russian economist writing during the early 20th century. Kondratieff noticed patterns of regular, structural change in the world-economy. These waves last 50 to 60 years and consist of two phases, a growth phase (the A-cycle) and a stagnation phase (the B-cycle).

Much of what drives these phases in modern economic world-systems comes from the cycles of exploitation and accumulation that we've already talked about. During the A-cycle, new products are created, markets expanded, labor employed, and the political and economic influence of core states moves into previously external areas—new geographic areas are brought into the periphery for labor and materials (imperialism). At 25 to 30 years into the A-cycle, profits begin to fall due to overproduction, decreasing commodity prices and increasing labor costs. In this B-cycle, the economy enters a deep recession. Eventually, the recession bottoms out and small businesses collapse, which leaves fewer firms and greater centralization of capital accumulation (quasi-monopolistic conditions), which, in turn, sets the stage for the next upswing in the cycle (A_2-cycle) and the next recession (B_2-cycle). Historically, these waves reach a crisis point approximately every 150 years. Each

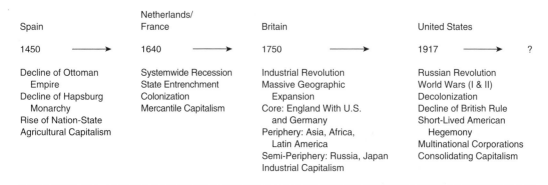

Figure 9.1 World-Systems Phases

wave has its own configuration of core and periphery states, with generally one dominant state, at least initially.

Wallerstein sees these waves as phases in the development of the world-system. Within each phase, three things occur: The dominant form of capitalism changes (agricultural → mercantilism → industrial → consolidation); there is a geographic expansion as the division of labor expands into external areas; and a particular configuration of core and periphery states emerges. There have been four such phases thus far in the world-system. In Figure 9.1, I've outlined the different phases and their movement through time. I've also noted some of the major issues and the hegemonic core nations for easy comparison. Wallerstein (2004) uses the term *hegemonic* to denote nations that for a certain period of time "were able to establish the rules of the game in the interstate system, to dominate the world-economy (in production, commerce, and finance), to get their way politically with a minimal use of military force (which however they had in goodly strength), and to formulate the cultural language with which one discussed the world" (p. 58).

I'm not going to go into much historic detail here. You can read Wallerstein's (1974, 1980, 1989) three-volume work for the specifics. But briefly, phase one occurred roughly between 1450 and 1640, which marks the transition from feudalism and world-empires to the nation-state. Both the Ottoman Empire and the Hapsburg dynasty began their decline in the 16th century. As the world-empires weakened, Western Europe and the nation-state emerged as the core, Spain and the Mediterranean declined into the semi-periphery, and northeastern Europe and the Americas became the periphery. During this time, the major form of capitalism was agricultural, which came about as an effect of technological development and ecological conditions in Europe.

The second phase lasted from 1640 to 1750 and was precipitated by a systemwide recession that lasted approximately 80 years. During this time, nations drew in, centralized, and attempted to control all facets of the market through mercantilism, the dominant form of capitalism in this phase. Mercantilism was designed to increase the power and wealth of the emerging nations through the accumulation of gold, favorable trade balances, and through foreign trading monopolies. These goals were achieved primarily through colonization (geographic expansion). As with the previous period, there was a great deal of struggle among the core nations, with a three-way conflict among the Netherlands, France, and England.

The third phase began with the Industrial Revolution. England quickly took the lead in this area. The last attempt by France to stop the spread of English power was Napoleon's continental blockade, which failed. Here capitalism was driven by industry, and it expanded geographically to cover the entire globe.

Wallerstein places the end of the third phase at the beginning of WWI and the beginning of the fourth phase at 1917 with the Russian Revolution. The Russian Revolution was driven by the lack of indigenous capital, continued resistance to industrializing from the agricultural sector, and the decay of military power and national status. Together these meant that "the Russian Revolution was essentially that of a semi-peripheral country whose internal balance of forces had been such that as of the late nineteenth century it began on a decline towards a peripheral status" (Wallerstein, 2000, p. 97). During this time, the British Empire receded, due to a number of factors including decolonization, and two states in particular vied for the core position: Germany and the United States. After the Second World War, the United States became the leading core nation, a position it enjoyed for two decades.

Hegemonic or leading states always have a limited life span. Becoming a core nation requires a state to focus on improving the conditions of production for capitalists; but staying hegemonic requires a state to invest in political and military might. Over time, other states become economically competitive and the leading state's economic power diminishes. In attempts to maintain its powerful position in the world-system, the hegemonic state will resort first to military threats and then to exercising its military power (note the increasing U.S. military intervention over the past 25 years). The "use of military power is not only the first sign of weakness but the source of further decline," as the capricious use of force creates resentment first in the world community and then in the state's home population as the cost of war increases taxation (Wallerstein, 2004, pp. 58–59).

Thus, the cost of hegemony is always high and it inevitably leads to the end of a state's position of power within the world-system. For the United States, the costs came from the Cold War with the USSR; competition with rising core nations, such as Japan, China, and an economically united and resurgent Western Europe; and such displays of military might as the Korean, Vietnam, Gulf, and Iraqi Wars. The decline of U.S. hegemony since the late 1960s has meant that capitalist freedom has actually increased, due to the relative size and power of global corporations. There are many multinational corporations now that are larger and more powerful than many nations. These new types of corporation "are able to maneuver against state bureaucracies whenever the national politicians become too responsive to internal worker pressures" (Wallerstein, 2000, p. 99). The overall health of world capitalism has also meant that the semi-periphery has increased in strength, facilitating growth into the core.

The Modern Crisis

There are several key points in time for the world-system, such as the Ottoman defeat in 1571, the Industrial Revolution around 1750, and the Russian Revolution in 1917. Each of these events signaled a transition from one capitalist regime to another. Wallerstein argues that one such event occurred in 1968, when revolutionary movements raced across the globe, involving China, West Germany, Poland, Italy, Japan, Vietnam, Czechoslovakia, Mexico, and the United States. So many nations were caught up in the mostly student-driven social movements that, collectively, they have been called the first world revolution. They were certainly powerful and extensive.

Wallerstein (1995) argues that these worldwide movements came out of the tension that has long existed between "two modernities—the modernity of technology and the modernity of liberation" (p. 472). Modernity grew out of the Enlightenment and positivistic philosophy. The Enlightenment refers to a period of European history from the late 1600s to the late 1700s that was characterized by the assumption that the knowledge of most worth is based on reason and observation (science) rather than faith (religion). Thinkers in the Enlightenment believed that, through reason, humans could control not only the physical universe but society as well. This was a positivist view of life as compared to fatalistic perspectives of religion. Humans could make a difference; humans could change their life course and not be subject to an impenetrable god. Part of this hope lay in technology: tools through which humanity could control the material universe and improve the physical standard of living. This hope was, of course, embodied in science and its offshoots, such as medicine.

The other pronounced hope of modernity concerned equality and was embodied in the nation-state; this is the second modernity. Many social scientists saw their place in this grand scheme of equality. For example, Harriett Martineau (1838/ 2003), one of the very first sociologists, argued that "every element of social life derives its importance from this great consideration . . . the relative amount of human happiness" (p. 25). For Martineau, the ultimate test of any society and its state is how well it measures up to this one great consideration. Nevertheless, modern society, especially in the United States, was founded on a contradiction: inalienable rights but only for a select group.

Wallerstein tells us that the upheavals of 1968 were directed at this contradiction and the failure of society to fulfill the hope of modernity: liberation for all. Students by and large rejected much of the benefits of technological development and proclaimed society had failed at the one thing that truly mattered: human freedom. The material benefits of technology and capitalism were seen as traps, things that had blinded people to the oppression of blacks, women, and all minorities. And this critique wasn't limited to technologically advanced societies. "In country after country of the so-called Third World, the populaces turned against the movements of the Old Left and charged fraud. . . . [The people of the world] had lost faith in their states as the agents of a modernity of liberation" (Wallerstein, 1995, p. 484). The 1968 movements in particular rejected American hegemony because of its emphasis on material wealth and hypocrisy in liberation.

In Wallerstein's scheme, the collapse of Communism was simply an extension of this revolt, one that most clearly pointed out the failure of state government to produce equality for all: "Even the most radical rhetoric was no guarantor of the modernity of liberation, and probably a poor guarantor of the modernity of technology" (p. 484). Interestingly, Wallerstein sees the collapse of Leninism as a disaster for world capitalism. Leninism had constrained the "dangerous classes," those groups oppressed through capitalist ideology and practice. Communism represented an alternative hope to the contradictions found in capitalist states. With the alternative hope gone, "the dangerous classes may now become truly dangerous once again. Politically, the world-system has become unstable" (p. 484).

Structurally, the upheavals of 1968 occurred at the beginning of a K-wave B-cycle. In other words, the world was standing at the brink of an economic downturn or stagnation, which lasted through the 1970s and 1980s. As we've seen, such B-cycles occur throughout the Kondratieff wave, but this one was particularly deep. The 20-year economic stagnation became an important political issue because of the prosperity of the preceding A-cycle. From 1945 to 1970, the world experienced more economic growth and prosperity than ever. Thus the economic downturn gave continued credence and extra political clout to worldwide social movements. Economically, the world-system responded to the downturn by attempting to roll back production costs by reducing pay scales, lowering taxes associated with the welfare state (education, medical benefits, retirement payments), and re-externalizing input costs (infrastructure, toxicity, raw materials). There was also a shift from the idea of developmentalism to globalization, which calls for the free flow of goods and capital through all nations.

However, while the world-system is putting effort into regaining the A-cycle, there are at least three structural problems hindering economic rebound. First, as we've noted, there are limits to exporting exploitation. Four hundred years of capitalism have depleted the world's supply of cheap labor. Every K-wave has brought continued geographic expansion, and it appears that we have reached the limit of that expansion. More and more of the world's workforce is using their political power to increase the share of surplus labor they receive. Inevitably, this will lead to a sharp increase in the costs of labor and production and a corresponding decrease in profit margins. Remember, capitalism is defined by continual accumulation. This worldwide shift, then, represents a critical point in the continuation of the current capitalist system.

Second, there is a squeeze on the middle classes. Typically, the middle classes are seen as the market base of a capitalist economy. And, as we've seen, a standard method of pulling out of a downturn is to increase the available spending money for the middle classes, either through tax breaks or through salary increases. This additional money spurs an increase in commodity purchases and subsequently in production and capital accumulation. However, this continual expanding of middle-class wages is becoming too much for firms and states to bear. One of two things must happen: Either these costs will be rolled back or they will not. If they are not reduced, "both states and enterprises will be in grave trouble and frequent bankruptcy" (Wallerstein, 1995, p. 485). If they are rolled back, "there will be significant political disaffection among precisely the strata that have provided the strongest support for the present world-system" (p. 485).

In the United States, indications are that the costs are being rolled back. Between 1967 and 2001, the income of the middle 20% of the population dropped from 17.3% to 14.6% of the total, while that of the upper 20% increased from 43.8% to 50.0%.

the modernity of technology on behalf of the modernity of liberation" (Wallerstein, 1995, p. 487).

The two structural signs that indicate we are in a time of chaotic transition are financial speculation and worldwide organization of social movements. There has been limited success in rolling back costs and reducing the press on profits, but not nearly what was needed or hoped for. As a result, capitalists have sought profit in the area of financial speculation rather than production. Many have taken great profits from this kind of speculation, but it also "renders the world-economy very volatile and subject to swings of currencies and of employment. It is in fact one of the signs of increasing chaos" (Wallerstein, 2004, p. 86).

On the political scene, since 1968 there has been a shift from movements for electoral changes to the "organization of a movement of movements" (Wallerstein, 2004, p. 86). Rather than national movements seeking change through voting within the system, radical groups are binding together internationally to seek change within the world-system. Wallerstein offers the World Social Forum (WSF) as an example. It is not itself an organization, but a virtual space for meetings among various militant groups seeking social change.

Another indicator of this political decentralization is the increase in terrorist attacks worldwide, such as the strike on the World Trade Center, September 11, 2001, and the bomb attacks on London, July 7, 2005. The terrorist groups themselves are decentralized, non-state entities, which makes conflict between a state like the United States and these entities difficult. Nation-states are particular kinds of entities defined by a number of factors, most importantly by territory, rational law, and a standing military. These factors and the political orientation they bring mean that nation-states are most efficient at confronting other nation states, ones that have specified territories, that legitimate rational law, and that have modern militaries. Almost everything about the terrorist groups that the United States is facing is antithetical to these qualities of the nation-state. The United States is a centralized state and the terrorists are decentralized groups. These differences in social structure and relation to physical place make it extremely difficult for the United States to engage the terrorists—there is no interface between the two—let alone defeat or make peace with them.

But more than that, the attacks of September 11 have energized politically right-wing groups in the United States. It has allowed them to cut ties with the political center and "to pursue a program centered around unilateral assertions by the United States of military strength combined with an attempt to undo the cultural evolution of the world-system that occurred after the world revolution of 1968 (particularly in the fields of race and sexuality)" (Wallerstein, 2004, p. 87). This, along with attempts to do away with many of the geopolitical structures set in place after 1945 (like the United Nations), has "threatened to worsen the already-increasing instability of the world-system" (p. 87).

These issues are the reason that Wallerstein (1999) talks about the demise of world-systems theory. Remember, world-systems theory is a critical perspective of global capitalism. As global capitalism fails, the insights that world-systems thinking can give us will become less significant. Yet in the remaining years of global capitalism, world-systems theory gives us a critical perspective for social involvement.

What will follow the 400-year reign of capitalism is uncertain. World-systems theory, as Wallerstein sees it, is meant to call our attention to thinking in structural time and cyclical processes; it is meant to lift our eyes from the mundane problems of our lives so that we can perceive the world-system in all its historical power to set the stage of our lives; it is intended to give us the critical perspective to have eyes to see and ears to hear the Marxian dynamics still at work within the capitalist system; and, finally, it is intended to spur us to action. Unlike C. Wright Mills (1956), Wallerstein is not saying that "great changes are beyond [our] control" (p. 3). Rather, he means that "fundamental change is possible . . . and this fact makes claims on our moral responsibility to act rationally, in good faith, and with strength to seek a better historical system" (Wallerstein, 1999, p. 3). Because the system is in a period of transition where "small inputs have large outputs" (1999, p. 1) and "every small action during this period is likely to have significant consequences" (Wallerstein, 2004, p. 77), we must make diligent efforts to understand what is going on; we must make choices about the direction in which we want the world to move; and we must bring our convictions into action, because it is our behaviors that will affect the system. In Wallerstein's (2004) words, "We can think of these three tasks as the intellectual, the moral, and the political tasks. They are different, but they are closely interlinked. None of us can opt out of any of these tasks. If we claim we do, we are merely making a hidden choice" (p. 90).

Summary

- Wallerstein sees his work more in terms of a type of analysis than a specific theory. His point is that it is the principles of analysis that drive the theorizing rather than the other way around. There are two main features of Wallerstein's perspective: globality and historicity. Globality conceptualizes the world in system terms, which cut across cultural and political boundaries. Historicity sees history in terms of structural time and cyclical time within the structures, rather than focusing on events, people, and linearity.
- In terms of theory, Wallerstein takes a Marxian approach. He focuses on the division of labor, exploitation, and the processes of accumulation and overproduction. In Marxian theory, exploitation is the chief source of profit. Thus, capitalists are intrinsically motivated to increase the level of exploitation. And since wages tend to go up as capitalist economies mature, reducing the level of exploitation and profit, there is therefore a constant tendency to export exploitation to nations that have a less developed capitalist economy, thus increasing the worldwide division of labor.
- Capitalist accumulation implies that capital is invested for the purpose of creating more capital, which in turn is invested in order to create more capital. In modern capitalism, this process of accumulation is augmented by the state. The state specifically bears the costs associated with transportation,

CHAPTER 10

Social Systems and Their Environments

Niklas Luhmann (1927–1998)

Photo: Courtesy of University of Bielefeld.

The last two chapters have addressed stratification, specifically gender and class inequality. The last three chapters of the book will also look at stratification issues—race, gender, and sexual inequality—but from a more radical point of view. All five of these chapters, and the theorists they represent, assume that stratification is a problem, one that can be solved. This idea of solving these kinds of issues is linked to what many see as the basic premise of modernity: reason and rationality. Human beings can rationally decide to make the world better.

But what if stratification and inequality aren't really problems? What if stratification can't be "solved"? What if the idea of inequality is actually an effect of the way we think about society and not really an issue at all? *What if inequality isn't really a part of society at all?* These are profound and disturbing questions. But what if the ideas in back of those questions are right? What would the ramifications be? Speaking just as a sociologist, I can tell you that one important ramification would be that most of sociology has been moving in the wrong direction for almost 200 years.

Obviously, I'm not starting this chapter off with these questions without purpose. I'm asking them because this is exactly what Niklas Luhmann wants us to consider: We may be wrong in the way we've been thinking about society—society isn't what most of us think it is, and stratification isn't what most of us think it is. And for the most part, sociology has been wrong about society for 200 years.

To get at this argument of Luhmann's, we have to start with a very basic question: What is society? We've touched on this topic a bit in the chapters on people in interaction. There, society is a generalized other or found in the ethnomethods of micro-organization or in networks of exchange. And we've looked at this issue in terms of cultural reality and structures of inequality. For those theorists, society exists as an objective structure, a thing that can be discovered and managed. However, Luhmann is going to challenge all our preconceptions of society, perhaps more than anyone else. What is society? And what are the ramifications of thinking about society differently?

The Essential Luhmann

Biography

Niklas Luhmann was born on December 8, 1927, in Lüneburg, Germany. He initially studied law and worked as a public administrator for over 10 years. On a work sabbatical in 1961, Luhmann studied Talcott Parsons's theories at Harvard. He began lecturing in 1962 at the University for Administrative Sciences in Speyer, Germany, and published his first of over 30 books in 1964, a study in formal organizations. In 1968, Luhmann took his first position as professor of sociology at the University of Bielefeld in Germany where he stayed until his retirement in 1993.

Passionate Curiosity

Luhmann's concern is sociology's big questions: What is society, and how does it work at a whole? His first exposure to the issue of society and how it functions came through Talcott Parsons. Luhmann was specifically concerned with modern society and argued vigorously with Jürgen Habermas about the differences between "rational modernity" and modern society as a complex system.

Keys to Knowing

organismic analogy, requisite needs, social evolution, social structures, cybernetic hierarchy of control, generalized media of exchange, equilibrium, open and closed systems, system environments, risk and complexity, co-evolution, autopoiesis, self-thematization, social systems, three processes of social evolution, segmental differentiation, stratified differentiation, functional differentiation, modern society, positive law, leftover vocabularies

---

Luhmann's Perspective: Thinking Systemically

Functionalism

Luhmann's perspective is heavily informed by functionalism and systems theory. **Functionalism** is principally based on the organismic analogy and the work of three theorists: Herbert Spencer, Émile Durkheim, and Talcott Parsons (see Allan, 2005a). The *organismic analogy* is a way of looking at society using organisms as a model. The fundamental idea taken from this analogy is that organisms have *requisite needs*. These needs push the organism to select and create internal structures in order to meet those needs. For example, in order for you to survive, you need oxygen and you get your oxygen from air. Because of that need and the way it is met, your body has a specific organ or structure—your lungs. Other organisms such as fish don't have lungs because they don't get oxygen from air. By analogy, the same is true for society: Different social structures meet specific needs.

The organismic analogy also implies evolution. Evolution basically occurs to enhance an organism's survivability. Generally speaking, the more complex an organism is, the greater will be its chances of surviving. Complexity is defined in terms of structural differentiation and specialization. Initial organisms were single-celled; that is, they only had one structure. As evolutionary processes continued, organisms became increasingly more complex. They developed different structures to meet specific needs, which, in turn, enhanced the organism's ability to survive. The human body, for example, is made up of many different kinds of structures (such as heart, lungs, liver, and bones) and many different subsystems (digestive system, respiratory system, nervous system, and so on).

The same is true for society. In fact, it is interesting to note that it was actually Herbert Spencer who coined the phrase "survival of the fittest," not Charles Darwin. As society evolves, it becomes more structurally differentiated and specialized. Generally speaking, social structures are made up of connections among sets of positions that form a network. The interrelated sets of positions in society are generally defined in terms of status positions, roles, and norms. These social and cultural elements create and manage the connections among people, and it is the connections that form the structure. Structural differentiation in society, then, is the process through which social networks break off from one another and become functionally specialized.

Social evolution, however, creates a problem. Spencer called it the *problem of coordination and control:* If social structures are distinct and specialized, how are their actions coordinated and controlled? Spencer generally argued that power centralized in the state provides the necessary control of diverse parts. Durkheim phrased the problem in terms of social integration. Parsons, drawing on Durkheim, came up with a different approach, one that is particularly interesting for us to consider in reference to Luhmann. Parsons's idea is called the *cybernetic hierarchy of control.* In the human body, cybernetics is the study of the autonomic nervous system. This system is formed by the brain and nervous system, and control is created through mechanical-electrical communication systems and devices. This kind of control is based on communication rather than power, and communication takes place through *generalized media of exchange.*

Think about it this way: The heart and lungs use blood as a medium of exchange. Blood circulates between the heart and lungs, exchanging carbon dioxide for oxygen. But what is the medium of exchange between the brain and the heart and lungs? How does your body communicate with itself generally, among all the diverse structures and systems? The human body uses electrical signals and the autonomic nervous system to control both the heart and lungs. In society, the generalized media is found in culture, which operates through the cybernetic hierarchy.

The organismic analogy implies one further issue for functionalism: equilibrium. *Equilibrium* is a state of balance between or among opposing forces or processes resulting in the absence of change. Most organic systems will tend toward equilibrium because it is the natural state of life. If we look at the physical or animal world, we see a great deal of overall stability. In fact, that is one of the problems with which evolutionists are faced: We don't see much change in our lifetime or even the life spans of many, many generations.

One of the reasons for this slowness is undoubtedly due to the fact that all things appear to exist within interrelated systems and subsystems. Because systems are interrelated, sudden change would probably bring chaos instead of order. In the same way, sudden change in one societal subsystem will tend to bring chaos, unless the change is countered with equal changes in other subsystems. In other words, unless social changes are met with equilibrating pressures, they will lead to the demise of that society.

There are, then, five defining features of structural-functionalism, as it comes to us in the Spencer-Durkheim-Parsons tradition:

1. Every system has requisite needs that must be met in order for that system to survive.

2. Specialized structures function to satisfy the needs of the system. Structures, functions, and the systemic whole are thus intrinsically related.

3. Specialization of structures occurs through the evolutionary process of differentiation.

4. Differentiation creates problems of coordination and control, which, in turn, create evolutionary pressures for the selection of integrating processes.

5. Integrating processes tend to keep the system in a state of equilibrium.

Systems Theory

Luhmann blends elements of functionalism with ideas from systems theory to form a new approach to understanding society. In this chapter, we're going to move back and forth between how classic functionalism sees things and how Luhmann's system theory is different. So we'll have examples and specific comparisons as we progress. Right now I'm simply interested in laying out the general principles of each perspective. As you'll see, there are some clear overlaps between structural-functionalism and systems theory. But there are also some important differences.

In brief, there are at least four qualities of a system. First, a system is made up of interrelated parts. Your car and your body are both examples of systems. In this way, functionalism is a systems theory. It's fundamentally concerned with the relationships among the parts and the parts with the whole. But, as you'll see, functionalism is a limited kind of systems theory.

The second characteristic of a system is that it exists in an environment. We are all somewhat familiar with this idea; we talk about computer environments and "the environment" (meaning ecological systems). In terms of their environments, systems can be more or less open. An important word of caution, however: A system cannot be completely open or closed. One of the major points of Luhmann's theory is that systems are formed by boundaries between it and the environment. A completely open system would have no boundaries and would thus be part of the environment and not a system. Totally closed systems are impossible as well. A completely closed system would have to be a perpetual motion machine, with no loss of energy. So we need to think about the relationship between systems and their environments as running on a continuum.

Basically, open systems take in information or energy from their environment and closed systems do not. A good example of an open system is your body. It takes in energy and information (food, air, sense data) and is thus directly influenced by the environment. If the environment changes too rapidly, your body will die. Relatively closed systems take in less information and energy. As a system, your car is more closed than your body. At this moment, the chances are good that your car is just sitting somewhere. It's inactive yet remains a system because its parts are related to one another. Your body, on the other hand, is never inactive. It is always taking in and processing information and energy from the environment.

Third, systems are dynamic. Of course this is a variable, but all systems involve processes. As we've already noted, systems take in or process energy and informa-tion. The energy can be in the form of food for organic systems or electricity for mechanical systems. Parsons actually touches on this issue a little in his cybernetic hierarchy of control. Energy moves up the hierarchy from the organic system and information moves down. However, there is more to this idea of process than the presence of energy and information.

Dynamic systems have feed-forward and feedback dynamics. The basic distinc-tion is the feedback of information. Can the system adjust to changes in the envi-ronment? If so, there is some kind of feedback process in place. Your car is a good example of a system without feedback mechanisms. It's primarily a feed-forward process. The system doesn't feed back information from the environment so that the car can make adjustments. Considering the advances in computer technologies, this is probably a limited example, but you get the point. Your body, on the other hand, has a number of feedback systems in place. For example, it self-adjusts to changes in the external temperature. "Cold-blooded" organisms do not have this feature.

The fourth defining characteristic of a system is that systems can be smart or dumb. Generally speaking, feedback systems are smart, but not always. In addition to feedback, a system must have a goal and explicit mechanisms in place to make adjustments based on incoming information and the system's goal. Obviously, your body is a smart system and your car a dumb system. But mechanical systems can be smart.

The heating and cooling system in your home is a good example of a smart mechanical system (Collins, 1988, pp. 49–50). It's smart because it has a thermostat. The thermostat has three important elements: a goal state (the temperature you set it at), an information mechanism (its ability to read the temperature in your home), and a control mechanism through which it turns the air conditioner or heater on and off. Smart systems such as the thermostat tend toward equilibrium. It balances out the forces of hot and cold through its control mechanism to keep your house at a comfortable 73 degrees, or whatever temperature it was set to.

As I mentioned, there are a number of differences between functionalism and systems theory. Below is a short list. As we'll see, Luhmann's concern is primarily focused on the first few issues. And, as we'll see by the conclusion of the chapter, the implications of using systems theory rather than structural-functionalism are significant.

- Systems theory pays attention to the relationship between the system and its environment; structural-functionalism generally does not.
- There are no requisite needs in systems theory; functionalism is defined by the delineation of such needs.
- Systems do not necessarily tend toward equilibrium; functionalism generally posits an equilibrated state. For systems theory, a state of equilibrium is a consequence of a system being smart.
- Systems theory is focused on processes; functional theory is focused on structures. Functional theory thus tends to reify its concepts and systems theory does not.

As we move through this chapter, keep the distinctive features of functionalism and systems theory in mind. We'll see how Luhmann blends elements of each to form his own brand of systems theory. In the end, Luhmann is going to argue that looking at society as a system has profound implications for the way we do sociology. We'll finish the chapter with a brief review of the gains of using systems theory.

Concepts and Theory: Self-Referencing Systems

Environments and Complexity

For Luhmann, the concepts of function and functional analysis no longer belong to the system itself, as with functionalism, but, rather, "*to the relationship between system and environment*. The final reference of all functional analyses lies in the difference between system and environment" (Luhmann, 1984/1995, p. 177, emphasis added). It is important to note that *system environments* are made up of other systems. Luhmann sees systems as interdependent and thus mutually constitutive. This is easy to illustrate: Did the collapse of the Soviet Union influence the United States? It obviously did. It changed the entire global environment for the United States and all other nations as well. For example, the idea of first, second, and third world countries is no longer viable. In other words, with the demise of the USSR, the concept of "third world countries" ceased to exist and the United States has become something different than a first world country (the other first world country was Soviet Russia) because such terms had become obsolete.

Thus, for Luhmann, the important beginning point for functional analysis isn't the system itself, but, rather, the boundary between the system and its environment. This idea has two implications. First, it means that systems are defined in terms of boundaries. A system exists only if it is different from its environment, that is, if there's some kind of boundary between it and the environment. This means that neither the system nor the environment is more important than the other—it's the relationship and boundary that are important. The second implication is that "this leads to a radical de-ontologizing of objects as such" (Luhmann, 1984/1995, p. 177). Remember the problem of reification we saw in Chapter 1? Luhmann is saying that shifting analysis to the boundary means that treating social structures as real objects becomes extremely difficult, because we cannot treat *difference* as an object or thing. Thus, in Luhmann's neofunctionalism there are no objective, social structures (like the kind we looked at in Chapters 6, 7, and 8).

That being said, Luhmann argues that *the boundary between system and environment is created by reducing complexity and risk.* Risk is defined by the relationship between the system and the environment. A system must maintain a boundary between itself and its environment, and the boundary must reduce the risk of the system being overwhelmed by the contingencies of the environment. System boundaries also reduce complexity. Again, this is by definition. Systems must be less complex than their environments because environments are composed of other systems. We should also note that reducing complexity and risk are related. Systems reduce the risk of being overwhelmed by their environment by reducing complexity. And, the reduction of risk and complexity are active issues, both of which are tied up with survivability. Systems survive in their environments because they reduce risks; they do this by reducing the complexity of the environment so that certain elements can be controlled. Let's use the lung and respiratory system example again: The lungs reduce the complexity of their specific environment (air) by extracting oxygen (a simpler compound).

Meaning and Social Systems

So far we've been talking at a very abstract level and it's time to bring it down to our primary concern: the social system. As we've seen, systems are defined in terms of boundaries that reduce risk and complexity in the environment. The social system evolved a very specific way of doing this. In fact, this evolution actually involved two systems: the social system and the psychic system. In other words, people and society need one another; society is impossible without people and people are impossible without society. At any point, the one is the necessary environment of the other. Luhmann (1984/1995) argues that this co-evolution came about because of a common achievement: "We call this evolutionary achievement 'meaning'" (p. 59).

Seeing communication and meaning as creating the boundary between the environment and the system is unique to Luhmann. While Durkheim never really considered system-boundary issues, both Spencer and Parsons did. However, what they saw were structures that negotiated the boundary rather than created it. Notice again the shift in primary interest: from internal structures that function to maintain the system to the boundary that creates the system. This shift is extremely important for us to understand Luhmann's theory, which is based on the essential feature of the social system, not on something derived from the system (like structures).

The social system is created through meaning, which is the elemental nature of human beings—meaning is what makes the human psyche and social system unique. In terms of the system, meaning and communication are the ways in which complexity and risk are reduced, thus producing the system boundary. In reducing risk and complexity, there are three central issues that social systems address: time, space, and symbols.

Humans have to address these issues differently from any other species. Because the social system is created through meaning, time, space, and symbols have endless horizons. In other words, for humans, time, space, and signification are all potentially infinite. Take time, for example: We not only can communicate with one another about the beginning of the universe through physics, but through religion we can talk about before and after the beginning. Physics and religion both tell us that time and space are related. The domain of God, for instance, is eternal (outside time) and omnipresent (outside space). Of course, the infinite possibilities of time and space are based on the ability of humans to use meaning, and meaning itself (symbols) must be held back from its endless possibilities. Thus, social systems are defined and produced when meanings are created that orient actors to a specific past, present, and future; that delineate certain spatial relationships; and that restrict the endless possibilities of symbolic worlds.

Reflexivity

One of the issues that meaning introduces is reflexivity or self-reference (we've seen this in Chapters 1 and 2, and we'll see it again in Chapter 12). Take a moment to think about our past and all the crazy things we humans have believed: The earth is flat; the earth is the center of the universe; gods live on Mount Olympus; the universe was created by the water god Nu and the sun god Atum; *hekura* live inside shaman and devour the souls of their enemies; it's the manifest destiny of the white man to dominate the world. The list is endless and that's the point—meaning refers only to itself. This, according to Luhmann (1984/1995), is "the fundamental law of self-reference" (p. 37).

Luhmann uses the term *autopoiesis* to talk about the issues surrounding self-reference. The word is made up of two Greek words: *auto* meaning self, and *poiesis* meaning creation. Autopoietic systems, then, are self-producing. The term originated in the work of two Chilean biologists, Humberto Maturana and Francisco Varela (1991). A clear example of an autopoietic system in biology is the cell. Biological cells are made up of biochemical components and bounded structures. In life, the cells use these components and structures to convert an external flow of energy and molecules into their own components and structures. Thus, the elements of the biological cell reproduce themselves. And, according to Luhmann, society does the same thing.

Communication systems, then, are self-referencing. Meaning systems are completely closed; they refer only to themselves. The reflexive nature of the social system implies three things. First, "only meaning can change meaning" (Luhmann, 1984/1995, p. 37). People exist in systems of meaning, and they make decisions that influence the social system in which they are working. Actors in this sense are free agents: Their decisions are not constrained. However, people can only use meaning to make decisions about meaning. This sounds circular because it is. If you make a decision about what courses to take or what major to have, you do it within the education system. Even if people change the meaning of something—as gays and lesbians are working to do with marriage—it is done within an already existing meaning system.

The second thing that self-referencing systems imply is that they must be continually remade. Specifically, the boundary between the system and the environment must be maintained. Think about an obviously human-made artifact dug up from a recently found ancient archeology site. In order to figure out what it is, the researchers must attempt to reconstruct the culture and society that it came from. Obviously, the meaning of the object isn't in the object itself. This example is clear to us because of our perception of time. As Luhmann puts it, "system events disappear from moment to moment and subsequent events can be produced only via the difference between system and environment" (p. 177).

The third implication is that societies self-thematize. *Self-thematization* is based on the idea that social systems can be reflective. That is, society can think about itself. Societal thematization is much like individual thematization. You undoubtedly would be able to give me a clear and coherent answer if I asked, who are you? There's a theme about you; it's your self-identity. It's the story line around which you organize ideas and experiences about who you are. This theme does a few things for you. First, it makes you different from everybody else (your environment); second, it gives you meaning (by reducing the complexity of your real experiences through continuous time and space); and third, it forms the basis upon which you make decisions (for example, you decided to come to school because you saw the decision as part of your understanding of who you are). Social themes function in the same way. Self-thematization implies that social systems are self-organizing: Using meaning, they organize their environmental boundary as well as the boundaries within the system (such as between religion and education).

Concepts and Theory: Social Evolution

Three Societal Systems

Luhmann gives us three different societal systems: interactional, social (society), and organizational. The foundation of all social systems is communication. As such, the basic unit of the societal system is the interaction, where two or more people meaningfully interrelate their actions: "As soon as any communication whatsoever takes place among individuals, social systems emerge" (Luhmann, 1982, p. 70). For Luhmann, fact-to-face encounters provide the opportunity for an interlocking relationship of action through symbolic communication; speaking to one another automatically sets up a boundary and this boundary reduces complexity from all possible communications. In addition, face-to-face communication is self-limiting in the sense that only one person generally talks at a time and only one topic can be dealt with at a time.

The limitations of the interactional system force movement to a system of another type. In other words, there has to be a communication system that can connect your face-to-face interaction with other interactions. Society, then, "*is the comprehensive system of all reciprocally accessible communicative actions*" (Luhmann, 1982, p. 73, emphasis original). Society coordinates communication with and among all possible actors missing from a single case (your interaction), and society regulates or systematizes through the principle of possible communication.

Society, then, is the meaning system that is capable of embracing a number of interactions. This is accomplished symbolically, through such things as language and self-thematization. To see how this works, let's set up three interactions at two different times. We'll say that the first set of interactions happened 600 years ago, with one interaction taking place in Tenochtitlan (the Aztec capital), one in York (England), and the third in Luoyang City (China). Each of these interactions would take place using different languages and different societal themes. However, all the interactions taking place within those cities would be meaningfully linked and thus constitute a society—a separate, bounded system.

Now let's move those interactions into our time. Since Tenochtitlan no longer exists, we'll use the Mexican city of Tecate, but both York and Luoyang still stand. Now, what are the differences? Are the interactions still as separate, thus constituting different social systems? Chances are good that the three interactions will still be in three different languages. But are there themes that link the interactions? Chances are better today that there are such themes. And the chances increase as we move to greater population centers, such as London, Mexico City, and Hong Kong. There are two strong themes that cut across large numbers of interactions: capitalism and democracy. Such themes prompt Luhmann (1982) to conclude, "Today, there is only one societal system—society is a world society" (p. 73). We will come back to this notion of a world society when we consider differentiation and modernity.

But for now I want you to notice how unique Luhmann's idea of society is. Using the ideas of system boundaries and meaning, Luhmann is able to give us a much more flexible and robust definition of society. This definition escapes the limitations of seeing society in terms of a territory, language, and state (the Weberian approach); the drawbacks of defining society in structural-functional terms (Spencer, Durkheim, and Parsons); the limitations of defining society on the basis of economic relations (the Marxian approach); and the restrictions of conceiving of society as a set of structures. In Luhmann's theory, society is almost organic, moving and changing as people redefine their meanings and change their interactions.

The third social system is organization. This system is "inserted" between the societal and interactional systems. The purpose of organizational systems is to sustain artificial behaviors for long periods of time in order to accomplish specific goals. The behaviors are "artificial" in the sense that they aren't directly motivated by either aesthetic values or moral demands. For example, most people working in a McDonald's kitchen do not do so for the work's intrinsic value.

In addition to motivating people to work by providing rewards—generally money in capitalist societies—there are two other methods of obtaining compliance: role specification, and entrance and exit rules. All organizations have explicit behavioral expectations of its members. For example, the university of which you are a part expects you to behave as a student; that's your role in the organization. Further, the university counts on you to internalize a good portion of these role expectations, such as values for academic honesty and scholastic truth. Of course, the specificity and commitment demanded vary by type of organization and position. If you are planning to go on to medical school, for example, the role of doctor is more highly specified, and the attitudinal and motivational expectations are greater, than your current role as student.

Organizations also have explicit entrance and exit rules that help manage members' commitment to work. Let's continue to use the university example to see how this works. The university that you are attending has demanding entrance rules. In order to become a student, you had to meet several criteria, such as GPA and SAT scores. Entrance rules such as these create investment in organizational roles. Many organizations also have exit rules. The university is again a good example: You must fulfill specific requirements concerning courses and quality of work to successfully exit the university. Note that these entrance and exit requirements, as well as the role expectations, carry over into your next organizational position. In other words, your documented performance in one organization sets the expectations in the next organization (your boss is going to expect you to think and act like a college graduate).

Evolutionary Processes

As we saw with classic functionalism, evolution means differentiation. Luhmann agrees. However, because Luhmann defines functional issues in terms of the relationship between a system and its environment, his concerns are different. We'll see the differences as we talk our way through his theory.

Generally speaking, evolution occurs through three processes: variation, selection, and stabilization. Systems increase the possibility of their survival by having the ability to create *variations*. There are two important qualities that define the social system's ability to create options. First, social systems, because they are based on meaning and communication, aren't limited to organic constraints. Humans now evolve symbolically, by dreaming imaginary worlds and bringing them into existence. Second, the "*capacity for evolutionary variation* is guaranteed because language always offers the option of saying 'no'" (Luhmann, 1982, p. 266, emphasis original). Luhmann argues that communication systems are based upon codes, which have a binary quality to them. And binaries always contain opposites, such as good/bad, male/female, and so on. Since every idea by definition contains its opposite, "we can communicate new, surprising, and unsettling messages, and will be understood" (p. 266).

The next evolutionary principle is *selection*. Luhmann talks about this in terms of the differences between language and media of communication. Language itself offers almost endless possibilities. Thus, I can say that "I'm a 6-foot tall rabbit with the ears of an elephant." But one of the things that selection does is to "de-realize" some of what is possible. I can *say* almost anything, but not everything I say will be understood, which is the basic requirement for communicative success and selection.

Communicative success, and thus selection, is governed by recognized media. A medium is a means of effecting or conveying something. When we are talking about a generalized medium of communication or exchange, we are referencing such things as the ideas and beliefs surrounding truth, love, money, political power, art, and so on. These are legitimate codes or discourses that we reference, which in turn give a statement intelligible space. Thus, the things that I can intelligibly say through a medium of "faith" might be very different from what will have communicative success through a medium of "political power."

The third process in evolutionary change is *stabilization,* and this requires the formation of a system—every social change must be systemically stabilized. The media that I mentioned above, faith and power, are understandable to you because they are already a part of a system of communication. Faith exists within the communicative system of religion and political power within the system of government.

Patterns of Differentiation

Sociocultural evolution occurs initially through separating the three systems we noted above—interaction systems, organizational systems, and societal systems. Thus, the greater the level of differentiation, the greater will be the independence of these systems. As differentiation between levels is achieved, social reality becomes more complex and the systems can assume separate functions and set themselves off from one another.

Differentiation also occurs within each system. Luhmann argues that differentiation within a system takes place through replication. Systems differentiate internally along the same path that they used to differentiate externally. In other words, systems replicate themselves internally. For instance, organizational systems will differentiate internally by proposing, selecting, and systematizing different organizational forms. "Differentiation is thus understood as a reflexive and recursive form of system building. It repeats the same mechanism, using it to amplify its own results" (Luhmann, 1982, pp. 230–231).

As I mentioned earlier, Luhmann asks us to look at systems and their environments. Part of what this means is that every differentiated subsystem has three references: (1) the external environment common to all subsystems, (2) its relation to other subsystems within the larger system, and (3) its relationship to itself. For example, each state within the United States has a common external environment (the federal government), is differently related to each of the other states (subsystems), and each state has its own unique configuration of state and local governments (relation to self).

The implication is that Luhmann's theory of evolutionary change is much more dynamic than the previous ones. Classic functionalism did not give sufficient weight to issues of environment. Luhmann, however, recognizes the movements between systems and their environments, both internally and externally. These multiple relationships in the long run tend to increase the level of differentiation and system building exponentially. In other words, complexity breeds complexity. Initial differentiation will be small. But because of the environmental relationships of systems to systems, differentiation in one system makes for a more complex environment for other systems, which then have to differentiate, which further adds to the complexity of the environment, which again prompts system differentiation.

Luhmann argues that there are three primary patterns of differentiation.

1. *Segmentary differentiation,* which differentiates society into equal and alike subsystems. In this case, a primitive society using kinship as its principal organizational form will tend to duplicate or extend kinship when differentiation is needed. This results in a system that is large but not very complex, which in the long run reduces the number and kind of variations that the system can produce (evolution is thus hampered).

2. *Stratification differentiation,* which differentiates society into unequal subsystems. The organization of society becomes hierarchical, with some subsystems having greater power or status than others. While segmentation only duplicates its systems, stratification creates diverse systems. This kind of differentiation does two things: It increases the number and diversity of possible variations and adaptive systems, and it creates pressures for increased communication and generalized media of exchange. A more abstract medium is needed to facilitate communication among different kinds of groups, and communication increases because of the diversity, both within and between strata.

3. *Functional differentiation,* which organizes communication around special functions to be fulfilled at the level of society. This is the type of differentiation with which classic functionalism was concerned. In functional differentiation, there are institutional fields that link up different organizations or subsystems. You are undoubtedly familiar with the names of these institutions, such as education, government, family, and so on. Let's use education for an illustration. "Education" is really a group of different organizations linked by a particular culture and communication system. In education there are school districts, university systems, textbook and journal publishers, organizations that produce chalk and blackboards, and so on. These organizations are functionally related to one another and to other institutional fields (through more abstract means of communication). Notice that creating these institutional domains produces entirely new environments, each with its own set of issues.

I've outlined Luhmann's ideas about system evolution and differentiation in Figure 10.1. In comparison to the actual world, the diagram is fairly simple. But it should work to give us an idea of how Luhmann thinks about societal evolution and provide us with at least a sense of the complexity involved. As you can see, the basis of society is the interaction. It forms a system against an environment, which I am noting with a thick line. In this "before differentiation" phase, all tasks are fulfilled through this one interactive grouping.

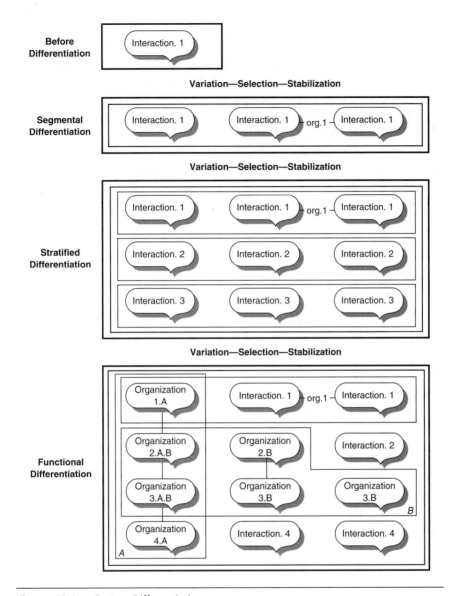

Figure 10.1   System Differentiation

Due to shifts in the environment, the interaction throws out different possibilities of variation, selection, and stabilization. Society then differentiates, but segmentally by duplicating itself (the interactions are all of the same type). The line between two interactions (Org.1) represents an organization. These organizations could take place among almost any of the interactions. This is still a simple social system, but notice that a new environment has been created. There is still the general, external environment, but now there is an internal one as well, formed by the communicative relations among the interactions and organizations.

Again, there is variation, selection, and stabilization, which move society to "stratified differentiation." Notice that now there are different types of interactions

(Interactions 1, 2, and 3), with any number of organizations occurring among them. These lines of like interactions constitute different social strata, as noted by the box surrounding each type. This is where society develops groups with different levels of resources to address different system concerns (in this case, there might be religious and political elites and lower-level economic workers). As a result of stratified differentiation, there is a corresponding increase in environmental complexity. There is still the external environment, but the internal social environment has expanded to include all interactions. In addition, there is also a social environment of like interactions (noted by the horizontal rectangles surrounding the different interactions).

With functional differentiation, things change dramatically. Notice that the social environment among like interactions still exists, as do the larger social system and the external boundary. Added to these are the institutional environments. I've noted two such environments, A and B, and connected all the organizations that constitute a functional domain. As you can see, some of the organizations are parts of two domains. An example of such an organization is Boeing, which is part of the airline industry as well as the military. The number and diversity of these crossover organizations increase as organizations include multinational, diversified corporations.

These patterns of differentiation roughly correspond to social evolution from simple to complex societies: Archaic societies are differentiated primarily through segmentation, high-culture societies through social strata or class, and modern societies by functional differentiation. What is unique about Luhmann's approach is that the complexity of society isn't gauged simply by the degree of differentiation, as classic functionalists would have it, but also by the incorporation of previous forms. Modern society is most complex because it incorporates the other two forms in addition to functional differentiation.

Before we look at modern societies, I want us to consider high-culture societies for a moment. In terms of historic epochs, high-culture societies occurred at different times in different locations. But when this social type occurred is less important than what it did socially. It's at this point in social evolution that society began to truly self-thematize. In segmental differentiation, the world and all it contained were relatively undifferentiated. That is, the physical, social, and spiritual worlds were seen as derivations of the same essence. High-culture societies, on the other hand, began to separate out different elements.

Religion, for example, moved from conceptualizing the world as infused with spiritual presence to seeing a complete separation between heaven and earth (see Weber, 1922/1963). Politically, society wrote its history as a story centered on itself and its purpose, achievements, and future. Within this political discourse were legitimating stories about those who could hold power and those who couldn't; these political stories increased the levels of diversity that society could coordinate. Economically, separating "nature" from humankind and the spiritual world allowed society to begin to exploit natural resources to their fullest. Luhmann (1982) characterizes this profound shift that came with high culture as the "gradual desocialization of nature" (p. 352); as nature is desocialized, "social reality becomes more complex" (p. 78). Thus, the very idea of society as a human institution is based on stratified differentiation: "In order for society to count as such, this and only this form of differentiation has to be recognized and accepted" (Luhmann, 1997, p. 68).

Modernity

To begin our discussion of modernity, we need to remind ourselves that Luhmann argues that modern society is global. Again, one of the theoretically powerful gains from using Luhmann's theory is the way it conceptualizes society. It clearly frees us from being constrained by territories or political systems. Political systems are certainly a type of social system, but in modernity they are a subsystem within the larger system of society. Keep this in mind as we review the general contours of modern social systems. These issues can be applied to any level, including the global social system.

As we noted earlier, initial differentiation occurs among the three different social systems: interactional, organizational, and societal. Differentiation among the three system levels also means that society can intervene in interactional or organizational systems without threatening its own survival. However, while they are differentiated, they are not disassociated. A total disjunction of the three levels is impossible since "all social action obviously takes place in society and is ultimately possible only in the form of interaction" (Luhmann, 1982, p. 79). This implies that the three subsystems are nested within and functionally dependent upon each other; this interdependency keeps conflict to a minimum. Conflict is also minimized because of the tendency toward piecemeal involvement. As societies differentiate, the individual becomes a nexus of diverse expectations that she or he cannot completely fulfill. You've undoubtedly experienced this already, with demands from parents, significant others, peers, work, school, religious organizations, and so on. The end result is that overt conflict between groups becomes increasingly difficult because people are spread too thinly among their emotional involvements.

In addition, differentiation allows members to be indifferent to the roles of others. Because our individual roles mean less to us in an environment of complex and irresolvable expectations, we don't have the strength of commitment to an identity to take exception with others. In other words, Luhmann sees modernity as a time where there is greater equality and acceptance. However, notice why this comes about. It isn't due to utopian beliefs about egalitarianism; it is simply due to the effects of the complex differentiation of society.

In terms of integration, one other important effect of differentiation remains—the autonomy of law: "The emergence of a functionally specific legal system appears to possess special significance as an achievement of social evolution: it is a condition for all further social evolution" (Luhmann, 1986/1989, pp. 129–130). Law represents specific communicative codes that persons and organizations can use to understand and regulate relationships. In this sense, the law is a special kind of medium of exchange within society. Society is based on the constraint of acts of communication, and law itself is a way of constraining communication. The separation of law as a function serves as a kind of evolutionary catalyst accelerating differentiation by facilitating communication across organizations and societies.

For social evolution and integration, it is important that law become able to reflexively produce its own themes, apart from social ideologies. This results in what Luhmann calls *positive law*. Before modernity, and even through the Enlightenment, law was seen as founded on a natural order or theocracy (rule by God). The well-known line from the Declaration of Independence—"We hold these truths to be self-evident, that all men are created equal, that they are endowed by their Creator with certain unalienable Rights"—is an expression of deistic natural law.

Natural law and divine law are absolute in every way: absolute truth and absolutely enforced. However, in the long run, this basis of legitimation hampers social evolution. America provides a case in point: Women and people of color were not included in the above quotation. If the United States had continued to believe in the natural or divine truth of the above statement, women, people of color, and so on would still not be considered citizens. Positive law rests solely upon the legislative and judicial decisions that make it law, and thus is more readily adaptable to a rapidly changing society. In contrast to natural law, the validity of positive law rests on the principle of variation: "It is the very alterability of law that is the foundation for its stability and its validity" (p. 94).

One of the things to notice about modern integration is that it is based on a generalized and flexible medium of exchange and that systems are integrated as a result of unintended consequences of differentiation: Modern society has no center and thus orderliness does not hinge upon structural (state, religion) or cultural (identity, nationalism) centrality. There's a way in which the mechanism for integration in highly differentiated societies is avoidance. What is not needed is a single set of shared value commitments—society has become too complex. Modern society is a highly abstract communicative network that simply defines vague and lax conditions for social compatibility, and it involves abstraction of and indifference to multiple aspects of the lives of individuals.

However, modern society isn't without its problems. An important issue of modern differentiation is that problems and issues can become displaced from the level of society to a subsystem. The result is that the societal subsystem responsible may not have the communicative tools (codes and themes) to deal with the issue. As an illustration, let's think about modern capitalism, the ecological environment, and peripheral (or underdeveloped) nations (see Luhmann, 1986/1989). The communicative scheme of capitalism is at odds with such humanitarian concerns as the poor or the environment. The profit motivation is intrinsic to capitalism—it is its defining theme. Without it, capitalism would not be capitalism.

Profit is accrued through expanding markets and increasing commodification, both of which are objectifying and amoral. We should expect, then, capitalist organizations to be oriented toward maximizing profit through expanding markets and commodification. Capitalist themes and codes are oriented toward bottom-line considerations, even in such global capitalist organizations as the World Trade Organization, which is assumed to have oversight over ecological issues. This thematic value for profit ought not to be seen as bad in and of itself. It is the result of capitalism's self-thematization. But it does tell us why ecological destruction is receiving so little real attention: The organizations that have been given the responsibility for it do not have the themes or communication systems to address it in any real way. (For another perspective concerning ecological destruction, see the *Web Byte, James O'Conner: Selling Nature.*)

Concepts and Theory: Changing Sociology's Question

Leftover Vocabularies

One of the implications of viewing society in terms of segmented, stratified, and functional differentiation is that stratified inequality becomes "functional." Now, that statement in itself is something that classic functionalists have said. For example, the classic functionalist argument is that class inequality is functional because it provides incentives for hard work and talent (Davis & Moore, 1945). This approach has been severely criticized by many, especially conflict sociologists. They see it as a way of justifying the status quo and class oppression.

However, Luhmann's argument is decidedly different from that of classic functionalism. As we've seen, stratification for Luhmann is part of the evolution of social forms. Stratification is a type of differentiation that occurred to increase the complexity of society so that it could reduce the complexity and risk of the environment through communication. Stratification is thus a boundary issue, not a parts–whole issue. This way of looking at differentiation implies that "stratification too presents us with a case of *system formation on the basis of equality*" (Luhmann, 1982, p. 263, emphasis original). Yes, stratification creates inequalities; that's the meaning of the term. But it principally generates equality because it creates a subsystem of equals that in turn produces more communication—stratification increases communication within the strata. Overall, then, communication is increased due to stratification, which is what "matters" to the social system.

Further, remember that Luhmann takes an evolutionary point of view. Generally, there is a move from simple to complex systems, with each major step in societal evolution being characterized by a specific type of differentiation: archaic societies—segmentation, high-culture societies—stratification, and modern societies—functional differentiation. While the previous forms continue, there is still segmentation in functionally differentiated societies; each social type is distinguished by a specific differentiation. Thus, stratification isn't what modern societies are based on, nor are we generally aware of stratification. In modern society, "the predominant relation is no longer a hierarchical one, but one of inclusion and exclusion, and this relates not to stratification but to functional differentiation" (Luhmann, 1997, p. 70).

What, then, are we to make of how conflict and critical theories think of inequality? Luhmann characterizes those theories as "leftover vocabularies." As we've seen, societies reflexively produce themes around which interaction and organization can take place. With increasing secularization, the theme of high-culture societies became "happiness." Thus, among the reasons for the creation of the United States was to guarantee "Life, Liberty, and the pursuit of Happiness." This theme soon shifted to the distinct themes of modernity: solidarity and equality.

Luhmann is thus arguing that, rather than analyses of how society works, the issues of solidarity and equality that sociology concerns itself with are actually part of political, ethical discourses. Further, even those societal self-thematizations are outdated. Think about how social integration occurs in functionally differentiated society—systemically speaking, solidarity and equality are no longer social concerns, particularly in a world-system society that is increasing in complexity. "Sociology may well see a task in correcting its own tradition and in shifting its attention from the outworn themes of stratification and compensatory social ideas to the more urgent external problems" (p. 74).

Society as System

As I said, Luhmann's understanding of the function of stratification is vastly different from that of classic functionalism. And, interestingly, he turns the tables on the conflict theorists who claim that functionalism is a value-driven perspective. As it turns out, it's the conflict and critical theorists who have provided the ideological fodder for modernity's discourse. But what do we gain by using Luhmann's systems theory?

To begin to answer this question, we need to revisit the idea of society as a system. Typically, society has been viewed as a territory that is controlled by a specific centralized state. Luhmann claims this approach is wrong-headed. Systems are defined by boundaries and boundaries by the reduction of environmental complexity and risk. Social systems do this through communication—communication and language are the basic tools through which human beings create boundaries and order. If Luhmann is right, then we have been misrecognizing society. Sociology has been thinking about itself more as an organization than a society.

Luhmann wants us to recognize that modern society is a world-system. Recall that modern societies are characterized as being functionally differentiated. Regional or national boundaries are not functional in Luhmann's (1997) sense: "They are political conventions, relevant for the *segmentary differentiation of the political subsystem* of the global society" (p. 72, emphasis added). Luhmann isn't saying that national or regional borders aren't boundaries; they are. But they are part of the segmentary system, not the functional system, which is becoming evolutionarily more dominant.

Recognizing society as a system changes sociology's questions. As I mentioned before, stratification and inequality are issues associated with differentiation and the evolution of society; they aren't a matter of "exploitation or suppression." Those are ideological ways of looking at the issue. Thinking in terms of exploitation and suppression isn't simply ideological; it is based on a specific concept of society: society as a thing, an entity that has a way of controlling itself. Luhmann's (1997) argument is that society is not a thing; it's a communication system that is evolving by becoming more complex. Society changes and evolves, "but cannot control itself" (p. 73). We can continue to make moral claims about society and stratification. "But who will hear these complaints and who can react to them, if the society is not in control of itself?" (p. 73).

Luhmann is asking us to consider looking at such problems differently. One of the ideas that we will see repeated in the next few chapters is the "failure of modernity." The hope of modernity was that rationality could make human existence better. Rationality could be used to control nature through technological advances and to ensure social equality, freedom, and happiness. Both of these modernities have failed to one degree or another. The unregulated use of technology is destroying the ecosystem, and social equality and happiness have always been at the expense of others. But maybe modernity didn't fail; maybe modernity could never have failed or succeeded, *because in this sense only organizations can fail or succeed.*

Luhmann's point is that systems don't exploit or suppress, they neglect. And systems don't decide to progress toward an ideological goal, they evolve. The issue, then, isn't the rational control of society; the issue is understanding how systems work. Thus, Luhmann confronts sociology with a decision to make. Oppression, poverty, and inequality continue; in fact, new oppressed groups continue to appear all the time, both "in" society and "out" of it. Perhaps the newest iteration is found in the terrorist attacks "on" the United States, Great Britain, and other "parts" of the "Western world." (All the words in "quotations" are terms based on the idea of society as an object or territorial identity—if we adapted Luhmann's perspective, this way of talking would fall out of our vocabulary.) Sociology can continue as it has, seeing society as an entity and attempting to find better ways of social integration around such ideas as happiness and equality. Or, sociology can see society for what it is: an evolving, complex system.

Complexity and Indeterminacy

The functional system requires a more abstract communicative system than the idea of nation-states can give. Earlier I gave you the example of three conversations. We saw that the interactions became more communicatively linked the closer we got to what we mean by modernity. This was one way for us to see that society is a global system, not just a national one. Another way to see this issue is to consider time. As we noted earlier, one of the complexities that social systems have to handle is time. This is done communicatively through social memory and speculation. Social memory (history) is selective, and only uses certain events to tell a meaningful story about society. And speculation (or oscillation, as Luhmann puts it) allows movement across the time barriers that memory erects. Using memory and speculation, time has been clearly used to produce segmented identities. That's why we have had different calendars with different years and events, such as the Chinese, Jewish, and Hindu calendars. But those segmented boundaries, though still present, are severely weakened in modernity.

The changes in time and space are a primary concern of theorists of modernity. We'll see this issue prominently in Anthony Giddens's theory (Chapter 12). For now we simply want to see that time has been relativized in modernity. As I said, every society clearly used social memory to form territorial identities; the link between these identities and time is seen in the diversity of calendars that used to be prevalent. And that's just the point: These calendars are no longer used in any significant way. The world has shifted to a universal clock and calendar. The calendar being used is the Gregorian calendar, a method of measuring time (and thus reducing complexity) that is originally based on Christianity. However, it is clear that the world is not using the calendar to understand itself as Christian. In fact, the traditional memory marking of "B.C." and "A.D." have given way to B.C.E. (Before Common Era) and C.E. (Common Era).

Luhmann's point is that as a result of system differentiation in modernity, the time boundary has become a global issue without segmented or stratified boundaries and is thus more reflexive. This strongly implies that the global system is evolving and becoming more complex. Rather than keeping memory in the past and speculation in the future, "the distinction of time re-enters itself" and we have to "live with the historical relativity of all cultural forms and with a lack of 'origins'" (Luhmann, 1997, p. 71). In other words, because time is being communicated and divided more abstractly, there isn't an origin (usually associated with national boundaries) as a reference point.

Again, let's use the calendars as an illustration. Each society that had a separate calendar linked its time-telling to a culturally significant event. The society reduced the complexity of time in that way, but did so in a segmented manner. Modernity is differentiated functionally, and *the shift in calendar use is indicative of society's evolution*—changes in the type and scope of differentiation. Universal time doesn't have an origin; it isn't marked by the birth of Christ or the cycles of the Chinese zodiac. Origins are now clearly seen as self-made, reflexive thematizations within a communicative system; "selectivity of reconfirmations and uncertainty of the future, are now unavoidable facts of social life" (Luhmann, 1997, p. 71). In other words, because of system differentiation, the modern world-society is complex and undetermined. System complexity means that sociology's causal explanations or policy plans are no longer possible. As Luhmann (1997) says, we are "in a phase of turbulent evolution without predictable outcome" (p. 76).

The Problem With Systems

The problems that sociology should be concerned with are system problems, not problems that are produced out of leftover vocabularies. And the theoretical problems that Luhmann faces us with are not slight. Let's take a look at one of them. Remember, systems are constituted by boundaries. If we as people want to pursue equality for all, then the problem isn't oppression and the solution isn't somehow forcing those in power to "do what's right." It isn't a problem of rationality. The problem is that it is the nature of systems to build boundaries. Intrinsic within systems, then, is a process of inclusion and exclusion, an inside and an outside. Further, evolution has developed systems "whose very complexity depends upon operational closure" (Luhmann, 1997, p. 73).

For this idea of system closure, Luhmann offers the example of the human brain. The human brain is arguably one of the most complex systems evolution has come up with, and it is radically shut off from its environment. Think about it this way: There isn't anything like your brain in the entire known universe. While your brain takes information in, its survival depends upon keeping all but select data out. The only thing like your brain is other human brains—and you really can't "get into their heads." Here's the bottom line: Your brain can be the most complex system we know of precisely because the boundary between it and its environment is so radical.

The problem is that society is exponentially increasing in complexity. What does that imply? It implies that society's boundary function is going to become more and more pronounced, which means its work of inclusion and exclusion will become more pronounced. As I said earlier, systems neglect but they don't exploit. Neglecting is part of how systems work. If something is excluded from the system, it is part of the environment and is either usable or unusable.

Under Luhmann's systems theory, the theoretical questions shift from issues of rationality to issues related to systems. Given the fact that the worldwide social system is growing in complexity, and given that as systems grow in complexity they become more bounded, the question becomes, how can we expect to include all kinds of concerns within the system? And what can we expect when we know that the very success of the system depends upon neglect? "*But this is a question and not an answer, and the question is meant to redirect sociological research*" (Luhmann, 1997, p. 75, emphasis added).

Summary

- Luhmann uses systems theory to argue that the primary issue for a social system is the boundary between it and its environment. Functional analysis should take place at this boundary point, not internal structural relations (as functionalism argues). This means that every system is different according to its environment and that at least part of a system's environment is made up of other systems. Systems theory pays attention to processes, not structures; systems vary by their degree of complexity.

- Environmental risk and complexity are reduced as systems differentiate and become more complex themselves. The risk and complexity that social systems must deal with revolve around time, space, and symbols. System differentiation is an evolutionary process that entails variation, selection, and stabilization. Human systems are created through communication; thus, social evolution involves variation in communication (provided by linguistic opposition), selection of new communicative forms (recognized media), and stabilization through creating new systems of communication.

- Because social systems create their environments through meaning, they are inherently self-referencing. The reflexivity of the social system implies that systems are self-organized (principally through self-thematization), self-produced, and are continuously remade.

- Societal systems have three distinct subsystems: the interaction system, the social system, and the organization system. Interaction systems are made up of face-to-face communication. Social systems (society) are comprehensive communication systems that link all reciprocally accessible communicative actions. Organizations are formal collectives with specific entrance and exit rules, roles, and goals that sustain artificial behaviors for long periods of time.

- Societal evolution differentiates among and within these subsystems through three different patterns: segmentation, stratification, and function. These patterns roughly correspond to increasing complexity in societal types: archaic, high-culture, and modern. Through each phase of differentiation, societies become more complex, first because each pattern increases the amount and diversity of communication and second because each evolutionary type contains the previous differentiation pattern. However, each evolutionary type has a dominant pattern: archaic—segmentation, high-culture—stratification, and modern—function.

- Differentiated societal systems integrate because society can influence interactional and organizational systems without endangering itself, because the different subsystems are nested, and because of positive law. Integration problems include interactional bottlenecks and problem dispersal to subsystems without the necessary communicative tools.

- Modern society is becoming an increasingly complex world-system. This has several implications for sociology and social policy. *Modern society is a complex communication system.* As such, the central problems for sociology and theory center around

  o The increasing complexity and thus indeterminacy of society
  o The necessity of system boundary work of environment/system, inclusion/exclusion
  o The tendency of complex systems to become closed

Building Your Theory Toolbox

Learning More—Primary Sources

See the following works by Niklas Luhmann:

- *The differentiation of society,* Columbia University Press, 1982.
- *Social systems,* Stanford University Press, 1995.
- *Observations on modernity,* Stanford University Press, 1998.
- *The reality of mass media,* Stanford University Press, 2000.

Learning More—Secondary Sources

- *Niklas Luhmann's modernity: The paradoxes of differentiation,* by William Rasch, Stanford University Press, 2000.

Check It Out

- *Web Byte—James O'Conner: Selling Nature*
- *Functionalism:* For a good historical overview of the development of functionalism, see Turner and Maryanski, *Functionalism,* Benjamin-Cummings, 1979.

- *Neofunctionalism:* Luhmann is generally categorized as a neo-functionalist. To find out more about this perspective, see J. C. Alexander's *Neofunctionalism and after,* Blackwell, 1998.
- *Complexity/Chaos theory:* For a good introduction in the social disciplines, see D. Byrne's *Complexity theory and the social sciences,* Routledge, 1998.

Seeing the World

- After reading and understanding this chapter, you should be able to answer the following questions (remember to answer them *theoretically*):
  o According to Luhmann, what makes a system a system?
  o In terms of reducing risk and complexity, what three things must human society address? Why are these specific problems for social systems?
  o Explain the three processes of evolution and how they work specifically in society.
  o Why is modern society more complex than either archaic or high-culture societies?
  o Through what processes are modern societies integrated? What problems are associated with modern societies?
  o In what ways does Luhmann's systems theory change the way we understand inequality?
  o What are the central issues with which sociological theory ought to be concerned?

Engaging the World

- How do improvements in communication technologies (such as the Internet and cell phones) affect the limitations imposed by face-to-face communication? Are there different boundary issues? Do such things as Internet chat rooms reduce or remove the limitations of turn-taking and topic?
- Given Luhmann's theory, how do you think worldwide ecological concerns can be addressed?

Weaving the Threads

- Compare and contrast Wallerstein's and Luhmann's ideas about a world-system. How can these two perspectives be reconciled?
- How do Wallerstein and Luhmann conceptualize the problem of ecological destruction? Given their theories, can environmental issues be successfully addressed? If so, how? If not, why not?
- Compare and contrast Luhmann's systems approach with the classic functional approach to differentiation. What does each perspective sensitize you to see?
- Each theorist that we've covered so far has a specific definition of society. Prepare a list of the different definitions. What do you make of the fact that there are so many different understandings of society? What do you think society is? Why?

This is a question that occupied most of our classical theorists. They lived and thought during a period of time when tremendous changes were taking place, and these people were driven to understand them. In comparison to traditional society, modernity is defined by the movement from small, local communities to large, urban settings; a high division of labor; high commodification and use of rational markets; and large-scale integration through nation-states. Rather than traditional and religious authority, modernity is characterized by individualism, rationality, bureaucracy, secularization, and alienation. In general, the defining moments of this modernity are the rise of nation-states, capitalism, mass democracy, science, urbanization, and mass media; the social movements that set the stage for modernity are the Renaissance, Enlightenment, Reformation, American and French Revolutions, and the Industrial Revolution.

Some elements of the Enlightenment were extremely important for setting the stage for modernity. The Enlightenment was a period of time in European history when Western ideas fundamentally changed. Perhaps the most profound shift is captured by the ideas of progress and rationalization. In traditional societies, religion was an important, if not the most important, social institution. Traditional, religious culture did not really contain the idea of progress. Things only changed if God revealed something new, which God wasn't in the habit of doing. The idea of progress came with a philosophy called positivism and the hope of human rationality.

In a nutshell, positivism is the belief that human beings can control their world. In positivism, life is no longer ruled by destiny, fate, or God. The hope of positivism is that humankind can take the helm by making rational choices based on scientific inquiry. Science assumes that the universe is empirical, operates according to law-like principles, and human beings can discover those laws and use them rationally. As we saw in Chapter 9, modernity is based on the application of these ideas in two realms: society and the physical world. The hope for the physical world is that humanity could control and use raw resources through technology to levels never before dreamed possible. The social hope was centered in the ideas of happiness, freedom, and equality. The physical and social/behavioral sciences were founded on these hopes.

One of the main questions of contemporary social theory has been whether or not we are still (or ever have been) modern in any or all of the ways I've just outlined. Our first two theorists in this section, Jürgen Habermas and Anthony Giddens, both believe that we are still living under modern conditions. Habermas in particular holds out the hope of modern reason and progress through specific kinds of speech acts and the revitalization of the public sphere. Giddens, on the other hand, while maintaining that society is still modern, argues that modernity by its nature is uncontrollable. For Giddens, the principal earmark of modernity is continual and accelerating change that is legitimated as progress. As such, modernity is like a runaway train: Humankind might be able to steer it to some degree but it also threatens to rush out of control and break itself apart.

There is an important point of comparison to bring out here. Habermas is decidedly modern in that he adheres to and theorizes about the hope of social progress through reason. Giddens, on the other hand, continues to think that society and technology might be controlled in some measure, but decidedly not through reason or emancipatory politics. In fact, Giddens argues that "sweet reason" and rational knowledge are part of the processes that produced a runaway world rather than a managed world. As we'll see, Giddens's idea of the possibility of social change is more akin to Luhmann's and Wallerstein's notions of complex systems and small inputs.

Our next two theorists are neither modern nor postmodern, per se. They are poststructuralists. Yet both theorists and poststructuralism itself contribute significant ideas to the modern/postmodern debate. With poststructuralism and postmodernism we have reached that point in the book that I told you about at the beginning. Remember, I began our journey with symbolic interaction for some important reasons: It set us up to think about meaning and the self. In poststructuralism and postmodernism, both meaning and self are seen as fragmenting to the point of nonexistence.

Poststructuralism argues that there are no structures, nor is human behavior determined or caused by anything. It denies that there is any firm base for behaviors, society, or reality. Using our building analogy, structure is nothing but smoke and mirrors. All we as humans have are discourse and text. Further, while there is no reality or meaning behind them, texts and forms of knowledge exert tremendous power over every aspect of our lives, primarily through discourse. The insidious part is that we control, limit, and objectify ourselves through discourse. In response, poststructuralism deconstructs the text or produces a counter-history of knowledge, which reveals the underlying and subtle political power found within all histories and discourses.

In some ways, the postmodern argument is like that of Giddens: Modernity contains dynamics that continually push for change. The difference is that postmodernists say that there has been a breach or rupture and we're no longer modern. Different theorists emphasize different social factors, but in my reading of the field there appear to be two effects that postmodernists are most concerned with: culture and the individual subject. Both are seen as simultaneously becoming more important and less real in some fashion.

Culture as culture is more important in postmodernity than in modernity. A major reason for this is the fact that postindustrial capitalism sells image more than product. Thus, "culture has necessarily expanded to the point where it has become virtually coextensive with the economy itself . . . as every material object and immaterial service becomes inseparably tractable sign and vendible commodity" (Anderson, 1998, p. 55). But at the same time, these processes make culture seem less real because it has little or no link to social networks.

Quite a bit of what postmodern capitalism sells are identity images. Thus, "as an older industrial order is churned up, traditional class formations have weakened, while segmented identities and localized groups, typically based on ethnic or sexual differences, multiply" (Anderson, 1998, p. 62). However, these segmented relationships "pull us in a myriad directions, inviting us to play such a variety of roles that the very concept of an 'authentic self' with knowable characteristics recedes from view. The fully saturated self becomes no self at all" (Gergen, 1991, p. 7). The saturated self that Gergen is talking about is one that is overwhelmed with media images.

CHAPTER 11

Modernity and Reason

Jürgen Habermas (1929–)

Photo: © Corbis.

S ocieties change—there's no doubt about that. But there is a great deal of debate about *how* societies change. In this book, for example, we've considered theories that say that society changes incrementally through each interaction or reproduction of human culture and reality. And we've looked at theories that argue that society changes as the result of massive systemic pressures. Jürgen Habermas confronts us with a different understanding altogether. Habermas argues that society can change because people choose to change it.

This vision of social change seems simple enough. In fact, it is the idea in back of democracy and a major reason why people vote. But we've already begun to see that this idea isn't as straightforward as it might seem. Niklas Luhmann and, to some extent, Immanuel Wallerstein have alerted us to the idea that the social system may be too complex for us to actually guide. And looking ahead in the book, Anthony Giddens (1990) similarly argues that modernity is like a runaway train, with no one at the helm; he asks, "Why has the generalising of 'sweet reason' not produced a world subject to our prediction and control?" (p. 151). In fact, from this point on in our book, reason and the audacious attempt of humans to control their world will be called into question again and again. The modern world is perhaps "not one in which the sureties of tradition and habit have been replaced by the certitude of rational knowledge" (Giddens, 1991, p. 3).

In response to such critiques, Erich Fromm (1955) pointed the way for critical theory when he said, "But all these facts are not strong enough to destroy faith in man's reason, good will and sanity. As long as we can think of other alternatives, we are not lost; as long as we can consult together and plan together, we can hope" (p. 363). In many ways, Habermas is one of the last, great modernists. Can we take control of society and move it to become better, more humane, and truly free? Can reason prevail in the face of the alienating forces of modernity? Habermas thinks so, and gives us theoretical reasons for our doing the same.

The Essential Habermas

Biography

Jürgen Habermas was born on June 18, 1929, in Düsseldorf, Germany. His teen years were spent under Nazi control, which undoubtedly gave Habermas his drive for freedom and democracy. His educational background is primarily in philosophy, but also includes German literature, history, and psychology. In 1956, Habermas took a position as Theodor Adorno's assistant at the Institute of Social Research in Frankfurt, which began his formal association with the Frankfurt School of critical thought. In 1961, Habermas took a professorship at the University of Heidelberg, but returned to Frankfurt in 1964 as a professor of philosophy and sociology. From 1971 to 1981, he worked as the director of the Max Planck Institute, where he began to formalize his theory of communicative action. In 1982, Habermas returned to the institute in Frankfurt, where he remained until his retirement in 1994.

Passionate Curiosity

Born out of the political oppression of Nazi Germany, Habermas was driven to produce a social theory of ethics that would not be based on political or economic power and would be universally inclusive. He is a critical theorist who sees humankind's hope of rational existence within the inherent processes of communication.

Keys to Knowing

critical theory, liberal capitalism, lifeworld, public sphere, organized capitalism, legitimation crisis, colonization of the lifeworld, colonization of the public sphere, communicative action, civil society

---

Habermas's Perspective: Critical Theory

Conflict theory began with Karl Marx and was significantly modified by Max Weber. Marx focused on the dynamics surrounding class, while Weber argued that the cross-cutting influences of class, status, and power significantly impact conflict and change in society. Weber also introduced a key element in stratification: legitimacy. But there is a further distinction between Marx and Weber. While Weber was disheartened and had grave concerns about modern life, especially related to bureaucracies and rationalization, he did not have the critical, revolutionary edge that Marx did. As a result, Marx has had a unique influence on contemporary social theory.

Marx spawned two distinct theoretical approaches. One approach focuses on conflict and class as general features of society. The intent with this more sociological approach is to analytically describe and explain conflict. The work of Erik Olin Wright (see the *Web Byte* introducing his work) is a clear example of such an

Concepts and Theory: Capitalism and Legitimation

Drawing on Karl Marx's theory of capitalism, Max Weber's ideas of the state and legitimation, Edmund Husserl's notion of the lifeworld, and Talcott Parsons's view of social systems, Habermas gives us a model of social evolution and modernity. You are by now generally familiar with Marx's theory of capitalism and Parsons's argument concerning the ways in which system components are integrated through a generalized media of exchange (Chapter 10). But let me take a moment to talk about Weber's and Husserl's contributions to Habermas's theory.

We came across the idea of legitimation in our review of Berger and Luckmann. There we saw that legitimations are stories through which human reality is given an institutionalized and moral basis. A Weberian concern with legitimacy is specifically, though not exclusively, focused on power, authority, and the state. In order for a system of domination like the state to work, people must believe in it. Part of the reason behind this need is the cost involved in the use of power. If people don't believe in authority to some degree, they will have to be forced to comply through coercive power. The use of coercive power requires high levels of external social control mechanisms, such as monitoring (you have to be able to watch and see if people are conforming) and force (because they won't do it willingly). To maintain a system of domination not based on legitimacy costs a great deal in terms of technology and manpower. In addition, people generally respond in the long run to the use of coercion by either rebelling or giving up—the end result is thus contrary to the desired goal.

In contrast to coercive power, authority implies the ability to require performance that is based upon the performer's belief in the rightness of the system, which is where legitimacy comes in. Legitimacy provides people with the moral basis for believing in the system. So, for example, your professor tells you that you will be taking a test in 2 weeks. And in 2 weeks you show up to take the test. No one has to force you; you simply do it because you believe in the right of the professor to give tests. And that's Weber's point: Social structures can function because of belief in a cultural system. The state, because it is almost exclusively defined in terms of power, is especially dependent upon legitimacy.

We've also seen the concept of the lifeworld with Berger and Luckmann as well. But, because it holds an important place in Habermas's theory, let me refresh your memory. The concept of **lifeworld** originally came from Edmund Husserl. Habermas uses it to refer to the individual's everyday life—the world as it is experienced immediately by the person, a world built upon culture and social relations, and thus filled with historically and socially specific meanings. The purpose of the lifeworld is to facilitate communication: to provide a common set of goals, practices, values, languages, and so on that allow people to interact, to continually weave their meanings, practices, and goals into a shared fabric of life.

Liberal Capitalism and the Hope of Modernity

Drawing from Marx and Weber, Habermas argues that there have been two phases of capitalism, liberal capitalism and organized capitalism. Each phase is defined by the changing relationship between capitalism and the state. In *liberal capitalism*, the state has little involvement with the economy. Capitalism is thus able to function without constraint. Liberal capitalism occurred during the beginning phases of capitalism and the nation-state.

Capitalism and the nation-state came into existence as part of sweeping changes that redefined Western Europe and eventually the world. Though they began much earlier, these changes coalesced in the 17th and 18th centuries. Prior to this time, the primary form of government in Europe was feudalism, brought to Europe by the Normans in 1066. Feudalism is based on land tenure and personal relationships. These relationships, and thus the land, were organized around the monarchy with a clear social division between royalty and peasants. Thus, the lifeworld of the everyday person in feudal Europe was one where personal obligations and one's relationship to the land were paramount. The everyday person was keenly aware of her or his obligations to the lord of the land (the origin of the word *landlord*). This was seen as a kind of familial relationship and fidelity was its chief goal. Notice something important here: People under feudalism were subjects of the monarchy, not citizens.

Capitalism came about out of an institutional field that included the state, Protestantism, and the Industrial Revolution. The nation-state was needed to provide the necessary uniform money system and strong legal codes concerning private property; the Protestant Reformation created a culture with strong values centered on individualism and the work ethic; and the Industrial Revolution gave to capitalism the level of exploitation it needed.

Habermas argues that together the nation-state and capitalism depoliticized class relations, proposed equality based on market competition, and contributed strongly to the emergence of the public sphere. The term *class* first came into the English language in the 17th century (see Williams, 1985, pp. 60–69). At that time, it had reference mainly to education; our use of *classic* and *classical* to refer to authoritative works of study came from this application. The true modern use of the term class came into existence between 1770 and 1840, a time period that corresponds to the Industrial Revolution as well as the French and American political revolutions.

Almost everything about society changed during this time, in particular the ideas of individual rights and accountability and the primacy of the economic system. The modern word class, then, carries with it the ideas that the individual's position is a product of the social system and that social position is made rather than inherited. "What was changing consciousness was not only increased individual mobility, which could be largely contained within the older terms, but the new sense of a society or a particular social system which actually created social division, including new kinds of divisions" (Williams, 1985, p. 62).

Thus, class is no longer a *political* issue, it is an *economic* one—class relations are no longer seen in terms of personal relations and family connections, but rather as the result of free market competition. Under capitalism and the civil liberties brought by the nation-state, all members of society are seen equally as citizens and economic competitors. Any differences among members in society are thus believed to come from economic competition and market forces, rather than birthright and personal relationships. Clearly, liberal capitalism brought momentous changes to the lifeworld: It became a world defined by democratic freedoms and responsibilities. Social relationships were no longer familial but rather legal and rational. The chief goal for the person in this lifeworld was full democratic participation. According to Habermas, the mechanism for this full participation is the public sphere.

The combination of the ideals of the Enlightenment, the transformation of government from feudalism to nation-state democracy, and the rise of capitalism created something that had never before existed: the public sphere. The **public sphere** is a space for democratic, public debate. Under feudalism, subjects could obviously complain about the monarchy and their way of life, and no doubt they did. But grumbling about a situation over which one has no control is vastly different than debating political points over which one is expected to exercise control. Remember, this was the first time Europe or the Americas had citizens, with rights and civic responsibilities; there was robust belief and hope in this new person, the citizen. The ideals of the Enlightenment indicated that this citizenry would be informed and completely engaged in the democratic process, and the public sphere is the place where this strong democracy could take place.

Habermas sees the public sphere as existing between a set of cultural institutions and practices on the one hand and state power on the other. The function of the public sphere is to mediate the concerns of private citizens and state interests. There are two principles of this public sphere: access to unlimited information and equal participation. The public sphere thus consists of cultural organizations such as journals and newspapers that distribute information to the people; it contains both political and commercial organizations where public discussion can take place, such as public assemblies, coffee shops, pubs, political clubs, and so forth. The goal of this public sphere is pragmatic consensus.

Thus, during liberal capitalism, the relationship between the state and capitalism can best be characterized as *laissez-faire*, which is French for "allow to do." The assumption undergirding this policy was that the individual will contribute most successfully to the good of the whole if left to her or his own aspirations. The place of government, then, should be as far away from capitalism as possible. In this way of thinking, capitalism represents the mechanism of equality, the place where the best are defined through successful competition rather than by family ties. During liberal capitalism, then, it was felt that the marketplace of capitalism had to be completely free from any interference so that the most successful could rise to the top. In this sense, faith in the "invisible hand" of market dynamics corresponded to the evolutionist belief in survival of the fittest and natural selection.

Organized Capitalism and the Legitimation Crisis

Such was the ideal world of capitalism and democracy coming out of the Enlightenment. The central orienting belief was progress; humankind was set free from the feudalistic bonds of monarchical government, and each individual would stand or fall based on her or his own efforts. In addition to economic pursuit, these efforts were to be focused on full democratic participation. Each citizen was to be fully and constantly immersed in education—education that came not only from schools but also through the public sphere. The hope of modernity was thus invested in each citizen and that person's full participation—people believed that rational discourse would lead to decisions made by reason and guided by egalitarianism.

Two economic issues changed the relationship between the economy and the state, which, in turn, had dramatic impacts on the lifeworld and public sphere. First, rather than producing equal competitors on an even playing field, free markets tend to create monopolies. Thus, by the end of the 19th and beginning of the 20th century, the United States' economy was essentially run by an elite group of businessmen who came to be called "robber barons." Perhaps the attitude of these capitalists is best captured by the phrase attributed to William H. Vanderbilt, a railroad tycoon: "The public be damned." These men emphasized efficiency through "Taylorism" (named after Frederick Taylor, the creator of scientific management) and economies of scale. The result was large-scale domination of markets. These monopolies weren't restricted to the market; they extended to "vertical integration" as well. With vertical integration, a company controls before-and-after manufacture supply lines. One example is Standard Oil, who at this time dominated the market, owned wells and refineries, and controlled the railroad system that moved its product to market.

The response of the U.S. government to widespread monopolization was to enact antitrust laws. The first legislation of this type in the United States was the Sherman Antitrust Act of 1890. In part the act reads, "Every contract, combination in the form of trust or otherwise, or conspiracy, in restraint of trade or commerce among the several States, or with foreign nations, is declared to be illegal. . . . Every person who shall monopolize, or attempt to monopolize, or combine or conspire with any other person or persons, to monopolize any part of the trade or commerce among the several States, or with foreign nations, shall be deemed guilty of a felony."

However, capitalists fought the act on constitutional grounds and the Supreme Court prevented the government from applying the law for a number of years. Eventually the Court decided for the government in 1904, and the Antitrust Act was used powerfully by both Presidents Theodore Roosevelt and William Taft. This regulatory power of the U.S. government was further extended under Woodrow Wilson's administration and the passing of the Clayton Antitrust Act in 1914.

The second economic issue that modified the economy's relationship with the state was economic fluctuations. As Karl Marx had indicated, capitalist economies are subject to periodic oscillations, with downturns becoming more and more harsh. By the late 1920s, the capitalist economic system went into severe decline, creating worldwide depression in the decade of the thirties. What came to be called "classic economics" fell out of favor and a myriad of competitors clamored to take its place. Eventually the ideas of John Maynard Keynes took hold and were explicated in his 1936 book, *The General Theory of Employment, Interest and Money*. His idea was simple, and reminiscent of Marx: Capitalism tends toward overproduction—the capacity of the system to produce and transport products is greater than the demand. Keynes's theory countered the then popular belief in the invisible hand of the market and argued that active government spending and management of the economy would reduce the power and magnitude of the business cycle.

Keynes's ideas initially influenced Franklin D. Roosevelt's belief that insufficent demand produced the depression, and after WWII Keynes's ideas were generally accepted. Governments began to keep statistics about the economy, expanded their control of capitalism, and increased spending in order to keep demand up. This new approach continued through the 1950s and 1960s. While the economic problems of the 1970s cast doubt upon Keynesian economics, new economic policies have continued to include some level of government spending and economic manipulation.

Thus, due to the tendency of completely free markets to produce monopolies and periodic fluctuations, the state became much more involved in the control of the economy. *Organized capitalism,* then, is a kind of capitalism where economic practices are controlled, governed, or organized by the state. According to Habermas, the change from liberal to organized capitalism, along with the general dynamics of capitalism (such as commodification, market expansion, advertising, and so on), have had three major effects.

First, there has been a shift in the kind and arena of crises. As we've seen, liberal capitalism suffered from economic crises. Under organized capitalism, however, the economy is managed by the state to one degree or another. This shift means that the crisis, when it hits, is a crisis for the state rather than the economy. It is specifically a legitimation crisis for the state and for people's belief in rationality.

There are two things going on here in the relationship between the state and the economy: The state is attempting to organize capitalism, and the state is employing scientific knowledge to do so. Together, these issues create crises of legitimation and rationality rather than simply economic disasters. Nevertheless, Habermas argues that the economy is the core problem: Capitalism has an intrinsic set of issues that continually create economic crises. However, due to the state's attempts to govern the economy, what the population experiences are ineffectual and disjointed responses from the state rather than economic crises. More significantly, in attempting to solve economic and social problems, the state increasingly depends upon scientific knowledge and technical control. This reliance on technical control changes the character of the problems from social or economic issues to technical ones.

Concepts and Theory: The Colonization of Democracy

The other two important effects concern the lifeworld and the public sphere. In our discussion of legitimation and rationality crises, we can begin to see the changes in the lifeworld. The lifeworld of liberal capitalism was constructed out of a culture that believed in progress through science and reason. In this lifeworld, the person was expected to be actively involved in the democratic process. However, the general malaise that grows out of the crisis of legitimation reduces people's motivation and the meaning they attach to social life.

Colonization of the Lifeworld

In addition, according to Habermas, the lifeworld is becoming increasingly colonized by the political and economic systems. To understand what Habermas means, we have to step back a little. As I've already noted, Habermas gives us a theory that involves social evolution. In general, social evolutionists argue that society progresses by becoming more complex: Structures and systems differentiate and become more specialized. The evolutionary argument is that this specialization and complexity produce a system that is more adaptable and better able to survive in a changing environment.

One of the problems that comes up in differentiated systems concerns coordination and control, or what Habermas refers to as "steering." You were introduced to this idea in Chapter 10. Remember, the problem is one of trying to guide social structures that have different values, roles, status positions, languages, and so forth. Differentiated social structures tend to go off in their own direction. We have seen that Talcott Parsons felt this problem was solved through generalized media of exchange. The idea of media is important, so let's consider it again for a minute. Merriam-Webster (2002) defines *medium* (media is plural) as "something through or by which something is accomplished, conveyed, or carried on." For example, language is a form of media: It's the principal medium through which communication is organized and carried out. Different social institutions or structures use different media. In education, for instance, it's knowledge and in government it's power. These are the instruments or media through which education and government are able to perform their functions.

For Parsons, the solution to the problem of social integration and steering is for the different social subsystems to create media that are general or abstract enough that all other institutions could use them as means of exchange. We can think about this like boundary crossings. Visualize a boundary between different social structures or subsystems, such as the economy and education. How can the boundary between economy and education be crossed? Or, using a different analogy, how can the economy and education talk to each other when they have different languages and values?

Habermas is specifically concerned with the boundaries between the lifeworld and the state and economy. In Habermas's terms, Parsons basically argues that the state and economy use power and money respectively as media of exchange with the lifeworld. If you think about this for a moment, it seems to make sense. You exist in your lifeworld, so what does the economy have that you want? You might start a list of all the cars, houses, and other commodities that you want, but what do they all boil down to? Money. And how does the economy entice you to leave your lifeworld and go to work? Money. So, money is the medium of exchange between the lifeworld and the economy. The same logic holds for the boundary between the lifeworld and the state: Power is what the state has and what induces us to interact with the state. However, Habermas (1981/1987) sees a problem:

> I want to argue against this—that in the areas of life that primarily fulfill functions of cultural reproduction, social integration, and socialization, mutual understanding cannot be replaced by media as the mechanism for coordinating action—that is, it *cannot be technicized*—though it can be expanded by technologies of communication and organizationally mediated—that is, it can be *rationalized*. (p. 267, emphasis original)

Habermas is arguing that there is something intrinsic about the lifeworld that cannot be reduced to media, such as money and power, "without sociopathological consequences" (p. 267). Let me give you an easy example from a different issue: having sex. Most people would agree that you cannot "technicize" this behavior using the medium of money without fundamentally changing the nature of it; there is a clear distinction between making love with your significant partner and having sex with a prostitute. Habermas is making the same kind of argument about humanity and communication in general. For him, the sphere of mutual understanding, the lifeworld, cannot be reduced to power and money without essentially changing it.

Yet Habermas isn't arguing that Parsons made a theoretical mistake. Parsons saw himself as an empiricist and merely sought to describe the social world. So in this sense, Parsons was right: There is something going on in modernity that tries to mediate the lifeworld. Habermas takes this idea from Parsons and argues that in imposing their media on the lifeworld, the state and economy are fundamentally changing it. The lifeworld, by definition, cannot be mediated through money or power without deeply altering it.

According to Habermas, the lifeworld is naturally achieved through consensus. This is basically the same thing that symbolic interactionists argue. Remember, interactions emerge and are achieved by individuals consciously and unconsciously negotiating meaning and action in face-to-face encounters. This negotiation, or consensus building, occurs chiefly through speech. Thus, using money or power fundamentally changes the lifeworld. In Habermas's (1981/1987) words, it is colonized: "The *mediatization* of the lifeworld assumes the form of a *colonization*" (p. 196, emphasis original).

This idea of the **colonization of the lifeworld** is perhaps one of Habermas's best known and most provocative concepts. Using Merriam-Webster (2002) again, a colony is "a body of people settled in a new territory, foreign and often distant, retaining ties with their motherland or parent state . . . as a means of facilitating established occupation and [governance] by the parent state." Habermas is arguing that the modern state and economic system (capitalism) have imposed their media upon the lifeworld. In this sense, money and power act just like a colony—they are means through which these distant social structures seek to occupy and dominate the local lifeworld of people.

Habermas (1981/1987, p. 356) argues that four factors in organized capitalism set the stage for the colonization of the lifeworld. First, the lifeworld is differentiated from the social systems. Historically, there was a closer association between the lifeworld and society; in fact, in the earliest societies they were *coextensive*; in other words, they overlapped to the degree that they were synonymous. As society increases in differentiation and complexity, the lifeworld becomes "decoupled" from institutional spheres. Second, the boundaries between the lifeworld and the different social subsystems become regulated through differentiated roles. Keep in mind that social roles are scripts for behavior. In traditional societies, most social roles were related to the family. So, for example, the eldest male would be the high priest—the family and religious positions would be filled and scripted by the same role. This kind of role homogeneity made the relationship between the lifeworld and society relatively nonproblematic, and, more importantly, it served to connect the two spheres.

Third, the rewards for workers in organized capitalism in terms of leisure time and expendable cash offset the demands of bureaucratic domination. "Wherever bourgeois law visibly underwrites the demands of the lifeworld against bureaucratic domination, it loses the ambivalence of realizing freedom at the cost of destructive side effects" (Habermas, 1981/1987, p. 361). And fourth, the state provides comprehensive welfare. Worker protection laws, social security, and so forth reduce the impact of exploitation and create a culture of entitlement where legal subjects pursue their individual interests and the "privatized hopes for self-actualization and self-determination are primarily located . . . in the roles of consumer and client" (p. 350).

For simplicity's sake, we can group the first two and last two items together. The first two factors are generally concerned with the effects of complex social environments. The more complex the social environment, due to structural differentiation, the greater will be the number and diversity of cultures and roles with which any individual will have to contend. This in turn dismantles the connections among the elements that comprise the lifeworld: culture, society, and personality.

The second two factors concern the effects of the state's position under organized capitalism. Under organized capitalism, the state protects the capitalist system, the capitalists, and the workers. In doing so, the state mitigates some of the issues that would otherwise produce social conflict and change. But perhaps more importantly, the state further individualizes the person. The roles of consumer and client, both associated with a climate of entitlement, overshadow the role of democratic citizen.

As a result of these factors, everything in organized capitalism that informs the lifeworld, such as culture and social positions, comes to be defined or at least influenced by money and power. Money and power have a certain logic or rationality to them. Weber talked about four distinct forms of rationality, two of which are pertinent here: instrumental and value rationality. Instrumental-rational action is behavior that is determined by pure means and ends calculation. For example, your action in coming to the university might be considered instrumentally rational if being here is a means to the goal of obtaining a good job or career. Value-rational behavior is action that is based upon one's values or morals. If there is no way you could get caught paying someone to write your term paper for you, then it would be instrumentally rational for you to do so. It would be the easiest way to achieve a desired end. However, if you don't do that because you believe it is dishonest, then your behavior is being guided by values or morals.

Value rationality is specifically tied to the lifeworld and instrumental rationality to the state and economy. Thus, a good deal of what happens when the lifeworld is colonized is the ever-increasing intrusion of instrumental rationality and the emptying of value rationality from the social system. The result is that "systemic mechanisms—for example, money—steer a social intercourse that has been largely disconnected from norms and values. . . . [And] norm-conformative attitudes and identity-forming social memberships are neither necessary nor possible" (Habermas, 1981/1987, p. 154).

In turn, people in this kind of modern social system come to value money and power; money and power are seen as the principal means of success and happiness. Money is used to purchase commodities that are in turn used to construct identities and impress other people. Rather than being a humanistic value, respect becomes something demanded rather than given, a ploy of power rather than a place of honor.

To see the significance of this, let's recall the ideal of the lifeworld of modernity. When the lifeworld changed in the move from traditional to modern society, it took on new priorities and importance. The lifeworld was ideally to be dominated by democratic freedoms and responsibilities and occupied by citizens fully engaged in reasoning out the ways to fulfill the goals of the Enlightenment—progress and equality—through communication and consensus building. As Habermas (1981/1987) says, "the burden of social integration [shifts] more and more from religiously anchored consensus to processes of consensus formation in language" (p. 180).

As you can see, using money or power as steering media in the lifeworld is the antithesis of open communication and consensus building. One of the results of this situation is that the lifeworld decouples from or becomes incidental to the social system, in terms of its integrative capacities. A lifeworld colonized by money and power cannot build consensus through reasoning and communication; people in this kind of lifeworld lose their sense of responsibility to the democratic ideals of the Enlightenment.

Colonization of the Public Sphere

This process is further aggravated by developments in the public sphere. As we've seen, the public sphere and its citizens came into existence with the advent of modernity. Citizens "are endowed by their Creator with certain unalienable Rights." This phrasing in the U.S. Declaration of Independence is interesting because it implies that these rights are moral rather then simply legal. There is a moral obligation to these rights that expresses itself in certain responsibilities:

> Whenever any Form of Government becomes destructive of these ends, it is the Right of the People to alter or to abolish it, and to institute new Government, laying its foundation on such principles and organizing its powers in such form, as to them shall seem most likely to effect their Safety and Happiness.

Thus the most immediate place for involvement for citizens is the public sphere. In that space between power on the one hand and free information on the other, citizens are meant to engage in communication and consensus formation. It is in that space that discussion and decisions about any "form of government" are to be made. However, the public sphere has been colonized in much the same way as the lifeworld. Specifically, the public sphere, which began in the 18th century with the growth of independent news sources and active places of public debate, transformed into something quite different in the 20th century. It became the place of public opinion—something that is measured through polls, used by politicians, and influenced by a mass media of entertainment.

There are two keys here. First, public opinion is something that is manufactured through social science. It's a statistic, not a public forum or debate that results in consensus. Recall what we saw earlier about how Habermas views the knowledge of science, even social science—its specific purpose is to control. Transforming consensus in the public sphere into a statistic makes controlling public sentiment much easier for politicians, both subjectively and objectively.

The second key issue I want us to see is the shift in news sources. Most of the venues through which we obtain our news and information today are motivated by profit. In other words, public news sources aren't primarily concerned with creating a democratic citizenry or with making available information that is socially significant. As such, information that is given out is packaged as entertainment most of the time. In a society like the United States, the consumers of mass media are more infatuated with "wicked weather" than the state of the homeless.

Concepts and Theory: Communicative Action and Civil Society

When we began our discussion, I mentioned that Habermas still holds out the promise of modernity. This hope is anchored in two arenas: speech communities and civil society. Both of these are rather straightforward proposals, though achieving them is difficult under the conditions created by organized capitalism, where the possibility and horizon of moral discourse are stunted.

Let's talk first about **ideal speech communities.** These communities or situations are the basis for ethical reasoning and occur under certain guidelines to communication. Before we get to those guidelines, we need to consider what Habermas calls communicative action: action with the intent to communicate. Habermas makes the point that all social action is based on communication. However, to understand Habermas's intent, it might be beneficial to consider something that looks like social communication but isn't. We can call this strategic speech. Strategic speech is associated with instrumental rationality, and it is thus endemic within the lifeworld of organized capitalism as well as the social system.

In this kind of talk, the goal is not to reach consensus or understanding, but rather for the speaker to achieve his or her own personal ends. For example, the stereotypical salesperson or "closer" isn't trying to reach consensus; she or he is trying to sell something (a more immediate example is the student explaining why she or he missed the test). In strategic talk, speech isn't being practiced simply as communication; communication is being *used* to achieve egocentric ends, which is contrary to the function of communication: "Reaching understanding is the inherent telos [ultimate end] of human speech. Naturally, speech and understanding are not related to one another as means to end" (Habermas, 1981/1984, p. 287).

Communicative action within an ideal speech situation is based upon some important assumptions. As we are reviewing these assumptions, keep in mind that Habermas is making the argument that communication itself holds the key and power to reasoned existence and emancipatory politics. Communication has intrinsic properties that form the basis of human connection and understanding. Habermas points out that every time we simply talk with someone, in every natural speech act, we assume that communication is possible. We also assume that it is possible to share intersubjective states. These two assumptions sound similar but are a bit different. Communication simply involves your assuming that your friend can understand the words you are saying. Sharing intersubjective states is deeper than this. With *intersubjectivity,* we assume that others can share a significant part of our inner world—our feelings, thoughts, convictions, and experiences.

A third assumption we make in speech acts is that there is a truth that exists apart from the individual speaker. In this part of speech, we are making validity claims. We claim that what we are saying has the strength of truth or rightness. This is an extremely important point for Habermas and forms the basis of discourse ethics and universal norms. All true communication is built upon and contains claims to validity, which inherently call for reason and reflection. Further, these claims assume validity is possible; that truth or rightness can exist independent of the individual, which implies the possibility of universal norms or morals; and that validity claims can be criticized, which implies that they are in some sense active and accountable to reason. Validity claims also facilitate intersubjectivity in that they create expectations in both parties. The speaker is expected to be responsible for the reasonableness of her or his statement, and the hearer is expected to accept or reject the validity of the statement and provide a reasonable basis for either.

These assumptions are basic to speech: We assume that we can communicate; we assume we can share intersubjective worlds; and we assume that valid statements are possible. What Habermas draws out from these basic assumptions of speech is that it is feasible to reasonably decide on collective action. This is a simple but profound point: Intrinsic to the way humans communicate is the hope of decisive collective action. It is possible for humanity to use talk in order to build consensus and make reasoned decisions about social action. This is both the promise and hope of modernity and the Enlightenment.

Ethical reason and substantive rationality are thus intrinsic to speech, but it isn't enough in terms of making a difference in organized capitalism. As with all critical theorists, Habermas has a praxis component. Praxis for Habermas is centered in communication and the creation of ideal speech situations. Here communication is a skill, one that as democratic citizens we need to cultivate in order to participate in the civil society. As we consider these points of the ideal speech community, notice how many of them have to do more with listening than with speaking. In an ideal speech situation,

- Every person who is competent to speak and act is allowed to partake in the conversation—full equality is granted and each person is seen as an equal source of legitimate or valid statements

- There is no sense of coercion; consensus is not forced; and there is no recourse to objective standings such as status, money, or power
- Anyone can introduce any topic; anyone can disagree with or question any topic; everyone is allowed to express opinions and feelings about all topics
- Each person strives to keep her or his speech free from ideology

Let me point out that this is an ideal against which all speech acts can be compared, and toward which all democratic communication must strive. The closer a community's speech comes to this ideal; the greater is the possibility of consensus and reasonable action.

> If we assume that the human species maintains itself through the socially coordinated activities of its members and that this coordination has to be established through communication . . . then the reproduction of the species also requires satisfying the conditions of a rationality that is inherent in communicative action. (Habermas, 1981/1984, p. 397)

Ideal speech communities are based upon and give rise to civil society. **Civil society** for Habermas is made up of voluntary associations, organizations, and social movements that are in touch with issues that evolve out of communicative action in the public sphere. In principle, civil society is independent of any social system, such as the state, the market, capitalism in general, family, or religion. Civil society, then, functions as a midpoint between the public sphere and social interactions. The elements of civil society provide a way through which the concerns developed in a robust speech community get expressed to society at large. One of the more important things civil society does is to continually challenge political and cultural organizations in order to keep intact the freedoms of speech, assembly, and press that are constitutionally guaranteed. Examples of elements of civil society include professional organizations, unions, charities, woman's organizations, advocacy groups, and so on.

Habermas gives us several conditions that must be met for a robust civil society to evolve and exist.

- It must develop within the context of liberal political culture, one that emphasizes equality for all, and an active and integrated lifeworld.
- Within the boundaries of the public sphere, men and women may obtain influence based on persuasion but cannot obtain political power.
- A civil society can exist only within a social system where the state's power is limited. The state in no way occupies the position of the social actor designed to bring all society under control. The state's power must be limited and political steering must by indirect and leave intact the internal operations of the institution or subsystem.

Overall, Habermas rekindles the social vision that was at the heart of modernity's birth. Modernity began in the fervor of the Enlightenment and held the hope that humanity could be the master of its own fate. There were two primary branches of this movement, one contained in science and the other in democratic society. In many ways, science has proven its worth through the massive technological developments that have occurred over the past 200 years or so. However, Habermas argues that the hope of democracy has run aground on the rocks of organized capitalism. In communicative action and civil society, he points the way to a fully involved citizenry reasoning out and charting their own course. But what Habermas gives us is an ideal—not in the sense of fantasy, but in the sense of an exemplar vision. In his theory, it is the goal toward which societies and citizens must strive if they are to fulfill the promise of modernity. Habermas, then, lays before us a challenge, "the big question of whether we could have had, or can now have, modernity without the less attractive features of capitalism and the bureaucratic nation-state" (Outhwaite, 2003, p. 231).

Summary

- Habermas's theory of modernity is in the tradition of the Frankfurt School of critical theory. His intent is to critique the current arrangements of capitalism and the state, while at the same time reestablishing the hope of the Enlightenment, that it is possible for human beings to guide their collective life through reason.
- Habermas argues that modernity has thus far been characterized by two forms of capitalism: liberal and organized. The principal difference between these two forms is the degree of state involvement. Under liberal capitalism, the relationship between the state and capitalism was one of *laissez-faire*. The state practiced a hands-off policy in the belief that the invisible hand of market competition would draw out the best in people and would result in true equality based on individual effort. However, *laissez-faire* capitalism produced two counter-results: the tendency toward monopolization and significant economic fluctuations due to overproduction. Both unanticipated results prompted greater state involvement and oversight of the capitalist system.
- Organized capitalism is characterized by active government spending and management of the economy. This involvement of the state in capitalism facilitates three distinct results, all of which weaken the possibility of achieving the social promise of modernity:

  1. A crisis of legitimation and rationality. Because the state is now involved in managing the economy, fluctuations, downturns, and other economic ills are perceived as problems with the state rather than the economy. When they occur, these problems threaten the legitimacy of the state in general. In addition, because the state uses social scientific methods to forecast and control economies, belief in rationality is put in jeopardy. These crises in turn reduce the levels of meaning and motivation felt by the citizenry.

2. The colonization of the lifeworld. The lifeworld is colonized by the state and economy, as the media of power and money replace communication and consensus as the chief values of the lifeworld.
3. The reduction of the public sphere to one of public opinion. This occurs principally as the media have shifted from information to entertainment value and as the state makes use of social scientific methods to measure and then control public opinion.

- However, Habermas argues that the hope of social progress and equality can be embraced once again through communicative action and a robust civil society. Communication is based upon several assumptions, the most important of which concern validity claims—these inherently call for reason and reflection. Together, such assumptions lead Habermas to conclude that the process of communication itself gives us warrant to believe it is possible to reach consensus and rationally guide our collective lives.

- Communicative action is also a practice. True communicative action occurs when full equality is granted and each person is seen as an equal source of legitimate or valid statements; objective standings such as status, money, or power are not used in anyway to persuade members; all topics may be introduced; and each person strives to keep her or his speech free from ideology.

- Communicative action results in and is based upon a robust civil society. Civil society is made up of mid-level voluntary associations, organizations, and social movements. Such organizations grow out of educated, rational, and critical communicative actions and become the medium through which the public sphere is revitalized. A civil society is most likely to develop under the following conditions: A liberal political culture is present that emphasizes education, communication, and equality; men and women are prevented from obtaining or using power in the public sphere; the state's power is limited.

Building Your Theory Toolbox

Learning More—Primary Sources

See the following works by Jürgen Habermas:

- *The theory of communicative action, vol. 1: Reason and the rationalization of society,* Beacon Press, 1984; *vol. 2: Lifeworld and system: A critique of functionalist reason,* Beacon Press, 1987.
- *The philosophical discourse of modernity: Twelve lectures,* MIT Press, 1990.
- *The structural transformation of the public sphere: An inquiry into a category of bourgeois society,* MIT Press, 1991.

Learning More—Secondary Sources

- *Habermas's critical theory of society,* by Jane Braaten, SUNY Press, 1991.
- *Habermas and the public sphere,* edited by Craig Calhoun, MIT Press, 1993.
- *Habermas: A critical introduction,* William Outhwaite, Stanford University Press, 1995.

Seeing the World

- After reading and understanding this chapter, you should be able to answer the following questions (remember to answer them *theoretically*):
  - o Define the Frankfurt School's critical theory and explain its view of knowledge and culture.
  - o Explain praxis and how it is associated with critical knowledge.
  - o Explain the differences between liberal and organized capitalism. Pay particular attention to the changing relations between the state and economy.
  - o Define the lifeworld and its purpose, and explain how it became colonized.
  - o Define the public sphere and explain how it came about, its purpose, and its colonization.
  - o What is communicative action and how does it form the basis of value-rational action?
  - o Define ideal speech situations (or communities) and explain how they give rise to civil society.
  - o What is civil society? What are the conditions under which it can survive? How is it important to a democratic society?

Engaging the World

- Using your favorite Internet search engine, look up "participatory democracy." How would Habermas's ideal speech community fit this model? Does the Internet provide greater possibilities for ideal speech situations to develop? How could Internet communities be linked to civil society?
- Racial, ethnic, gender, sexual identity, and religious groups have all been and are being disenfranchised in modern society. How does the ideal speech situation "enfranchise" these groups? In other words, how does the ideal speech situation do away with the possibility of disenfranchised groups?
- What social group do you belong to that most nearly approximates the ideal speech community?
- How can you begin your own praxis?

CHAPTER 12

The Juggernaut of Modernity

Anthony Giddens (1938–)

Photo: Courtesy of Anthony Giddens.

Have you ever ridden a rollercoaster? One of the things that makes riding a rollercoaster fun is the way danger and security are mixed together. We wouldn't ride the rollercoaster if we didn't believe it was safe; but the rollercoaster wouldn't be fun if we didn't have a sense of danger. When we slam into the curves and plummet over a hundred feet down, we feel the possibility of death, but it is tempered by our sense of trust in the machine and the experts who built it. Anthony Giddens pictures modernity in much the same way, but with some important differences.

According to Giddens (1990), modernity is a juggernaut, "a runaway engine of enormous power which, collectively as human beings, we can drive to some extent but which also threatens to rush out of our control and which could rend itself asunder" (p. 139). The word *juggernaut* comes from the Hindi word, *Jagannātha*, which refers to a representation of the god Vishnu or Krishna—the lord of the universe. Every year the god's image would be paraded down the streets amid crowds of the faithful, dancing and playing drums and cymbals. It's thought that at times believers would throw themselves under the wheels of the massive cart, to be crushed to death in a bid for early salvation. A juggernaut, then, is an irresistible force that demands blind devotion and sacrifice.

This image of an irresistible force conjures up the thrilling ride of the rollercoaster, with its twin sensations of trust and danger, but the juggernaut of modernity isn't as controllable or predictable as a rollercoaster. Here we can see a chief difference between Giddens and Habermas: For Habermas, rational control is central to modernity and imminently possible; but for Giddens, modernity is almost by definition out of control. The intent of modernity is progress—but the *effect* of modernity is the creation of mechanisms and processes that become a runaway engine of change. And we, like the devotees of Jagannātha, are drawn to modernity's power and promise.

> The ride is by no means wholly unpleasant or unrewarding; it can often be exhilarating and charged with hopeful anticipation. But, so long as the institutions of modernity endure, we shall never be able to control completely either the path or the pace of the journey. In turn, we shall never be able to feel entirely secure, because the terrain across which it runs is fraught with risks of high consequence. (Giddens, 1990, p. 139)

The Essential Giddens

Biography

Anthony Giddens was born January 18, 1938, in Edmonton, England. He received his undergraduate degree with honors from Hull University in 1959, studying sociology and psychology. Giddens did his master's work at the London School of Economics, finishing his thesis on the sociology of sport in 1961. From then until the early 1970s, Giddens lectured at various universities including the University of Leicester, Simon Fraser University, the University of California at Los Angeles, and Cambridge. Giddens finished his doctoral work at Cambridge in 1976. He remained at Cambridge through 1996, during which time he served as dean of Social and Political Sciences. In 1997, Giddens was appointed director of the London School of Economics and Political Science. Giddens is the author of some 34 books that have been translated into well over 20 languages. Giddens is also a member of the Advisory Council of the Institute for Public Policy Research (London, England) and has served as advisor to British prime minister Tony Blair.

Passionate Curiosity

Giddens is a political sociologist, driven by both political questions and political involvement. While his early work certainly contained a typical Marxian interest in class, his later work is much more concerned with the political ramifications of globalization and what he characterizes as the juggernaut of modernity or the runaway world. Given the juggernaut of modernity, he asks, how are interactions and behaviors patterned over time? How can people become politically involved? In order to answer those questions, Giddens must first understand the essence of society. In this, Giddens seeks an ontology of the social world: What kinds of things go into the making of society? Precisely *how* does it exist?

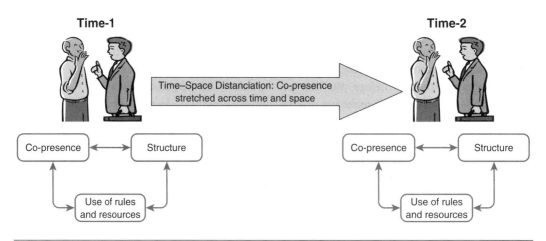

Figure 12.1 Duality of Structure

Modalities of Structuration

Notice the large arrow linking the two sets of interactions in Figure 12.1. The arrow indicates how behaviors and encounters are patterned over time. Giddens (1986) rephrases the problem of patterning behaviors in terms of time–space distanciation: "The fundamental question of social theory . . . is to explicate how the limitations of individual 'presence' are transcended by the 'stretching' of social relations across time and space" (p. 35). The idea of **time–space distanciation** refers to the ways in which physical co-presence is stretched through time and space. This is a fairly unique and graphic way of thinking about patterning behaviors. We can think of Giddens's idea as an analogy: If you've ever played with Silly Putty or bubble gum by stretching it out, then you can see what he is talking about. What this analogy implies is that the interactions at Time-1 and Time-2 appear patterned because they are made out of the same materials that are stretched out over time and space.

How this stretching out of time and space happens is Giddens's fundamental question. His answer is found in his idea of modalities of structuration. The word *modality* is related to the word *mode,* which refers to a form or pattern of expression, as in someone's mode of dress or behavior. For example, in writing this book, I'm currently in my academic mode. **Modalities of structuration,** then, are ways in which rules and resources are knowingly used by people in interactions.

I've pictured a bit of what Giddens is getting at in Figure 12.2. Notice that there are three elements in the circle: social practices, modalities, and structures. Modalities of structuration are ways in which structure and practice (or agency) are expressed. I've indicated that relationship by the use of overlapping diamonds. In a loose way, we can think of structures as the music itself; the modalities as the mode of reproduction, as in analog or digital; and the social practices as the musician. As you can see, Giddens gives us *three modalities or modes of expression* (interpretive

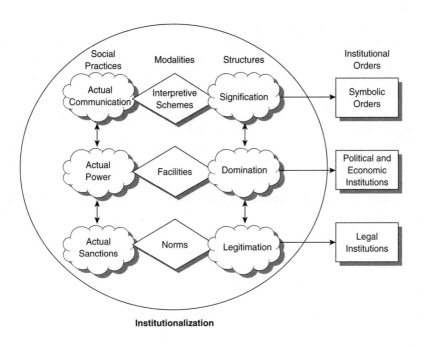

Social Practices | Modalities | Structures | Institutional Orders

Actual Communication — Interpretive Schemes — Signification → Symbolic Orders

Actual Power — Facilities — Domination → Political and Economic Institutions

Actual Sanctions — Norms — Legitimation → Legal Institutions

Institutionalization

Figure 12.2   Modalities of Structuration

schemes, facilities, and norms), corresponding on the one hand to three social practices (communication, power, and sanctions), and on the other to structures (signification, domination, and legitimation).

This isn't as complicated as it might seem. Let's use the example of you talking to your professor in class. Let's say that in this conversation you refuse to take the test that she or he has scheduled. The professor reacts by telling you that you will fail the course if you don't take the test. What just happened? You can break it down using Giddens's modalities of structuration, following the model in Figure 12.2.

First, there were actual social practices that involved communication and sanctions. Your communication was interpreted using a scheme that both you and your professor know. For convenience sake, let's call this scheme "meanings in educational settings." You know this interpretive scheme because it is part of the general signification structure of this society at the beginning of the 21st century. Second, the professor invoked sanctions based on norms of classroom behavior. Again, you both know these norms because they are part of the legitimation structure of this society.

I'm sure you were able to follow this discussion through the model without any difficulty. And I'm also sure that you didn't have any trouble with any of the phrases I used in the above explanation. Things like "society at the beginning of the 21st century" sound reasonable and familiar. But remember, the first principle of structuration theory is duality, not dualism. So when we use the terms society or structure we are not talking about something separate from social practices. Like our sentence illustration, the actual conversation between you and the professor, and the interpretive schemes and the structure of signification, all come into existence at the same moment. Apart from signification and interpretation, communication can't exist; and likewise, without actual communication, interpretation and communication can't exist. Obviously, the same is true for the sanctions that the professor invoked.

In terms of Giddens's definition of structure as rules and resources, signification and legitimation are more closely tied to rules and domination is more linked to resources (facilities). *Domination* is expressed as actual power through the facilities of authoritative and allocative resources. For example, part of the way the actual power of the university over me as a professor is expressed is through the facilities of classroom space, computer and Internet access, and so forth. By encircling all these elements together and by using two-headed arrows, I'm indicating that all these processes—the social practices, modalities, and structures—are reflexive and recursive. That is, they mutually and continuously influence one another.

Part of what I want you to see in Figure 12.2 is the connectedness of all social practices, modalities, and structures. They are all tied up together and expressed and produced in the same moment. Further, the recursive and interpenetrated nature of these facets of social life are what Giddens means by *institutionalization,* or the stretching out of co-presence across time and space. Remember, human life is ongoing. The process that I've placed a circle around in Figure 12.2 works like a ball that just keeps rolling downhill. It is this continuity of recursive practices and structures that stretches interactions across time and space. Comparing Figure 12.1 and Figure 12.2, we can think of the two men talking as moments in which we stopped the ball and looked inside. The arrow between the two sets of interactions depicts the movement of the ball between those two moments.

There's one other thing that we need to notice from Figure 12.2. All of this action of institutionalization results in different *institutional orders.* While there are some terms in there that look familiar, like "economic institutions," they aren't the same as we usually think of them. Many sociologists think of institutions as substantive and distinct (or "differentiated," in functionalist terms). In other words, most sociologists treat institutions as if they are real, separate objects with independent effects. However, Giddens is saying that institutions don't exist as substantive things or objects and they aren't truly separate and distinct.

Notice that the different institutional orders are all made from the same fabric; it's just cut or put together differently in each case. They all draw from the same structures (rules and resources) of signification (S), domination (D), and legitimation (L), but emphasizing one of the elements over the others produces different kinds of institutional orders. The way Figure 12.2 is mapped allows us to work out the different orders. The structure that is most closely associated with the institutional order is the most important.

Thus, in the case of symbolic institutional orders (political ideology is a good example), the order is drawn as S-D-L. Think about a political ideology like democracy: It is primarily based on signification and meaning with domination and legitimation backing the meanings and symbols. In the case of legal institutional orders, the arrangement is reversed from that of symbolic orders. A legal order is drawn as L-D-S. Its primary structural source is legitimation, closely followed by domination and, further back, signification. The other two institutional orders that Giddens gives us are drawn similar to one another but with different kinds of domination. Thus, political institutions are D (authoritative)-S-L and economic institutions are drawn as D (allocative)-S-L.

Giddens's Perspective: Reflexive Actors

Levels of Awareness

Giddens argues that there are three important things going on in interactions: reflexive monitoring of action, rationalization of action, and motivation for action. Giddens thinks of these tasks as being "stratified," or as having different levels of awareness. The behavior that is most conscious is *reflexive monitoring*. In order to interact with one another, people must watch the behaviors of other people, monitor the flow of the conversation, and keep track of their own actions. As part of this routine accomplishment, we can also provide reasons for what we do; that is, we can provide a rationalization of our own actions.

In talking about rationalization of action, Giddens makes a distinction between discursive and practical consciousness. The word *discursive* is related to discourse or conversation. But it has a deeper meaning as well: It's a discourse marked by analytical reasoning. So **discursive consciousness** refers to the ability to give a reasoned verbal account of our actions. It's what we know and can express about social practices and situations. This consciousness is clearly linked to reflexive monitoring of the encounter and the rationalization of action—discursive consciousness is our awareness of these two.

Practical consciousness refers to the knowledge that we have about how to exist and behave socially. However, people can't verbally express this knowledge. Social situations and practices are extremely complex, according to Giddens, and they thus require a vast and nuanced base of knowledge, and we have to act more by intuition than by rational thought. This idea isn't as difficult as it might seem. We can think of the ability to perform an opening ritual ("Hey, how's it going?") as part of this practical consciousness. People know *how* to perform an opening ritual, but most people can't rationally explain *why* they do it.

There's an important point to note about discursive and practical consciousnesses: They aren't necessarily linked. At first glance, it might appear that discursive consciousness is our ability to explain what practical consciousness tells us to do. But notice what I said above about practical consciousness: "People can't verbally express this knowledge." So discursive consciousness (the explanation) isn't necessarily associated in any real way with practical consciousness (the actions). We know how to act and we know how to explain our action, but both of these issues are part of the social interaction, not part of the unconscious motivations of the actor.

Unconscious Motivation

Practical consciousness is bound up with the production of routine. It's like driving a car or riding a bicycle; most of what is involved is done out of habit or practical consciousness. In the same way, most of what we do socially on a daily basis is routine. **Routinization** "is a fundamental concept in structuration theory" (Giddens 1986, p. xxii) and refers to the process through which the activities of day-to-day life become habitual and taken-for-granted. Routinization, then, is a primary way in which face-to-face interactions are stretched across time and space (Figure 12.1). Or put another way, routinization is one of the main ways through which the modalities of structuration are institutionalized (Figure 12.2). Part of the way we routinize activities is through **regionalization,** which is the zoning of time and space in relation to routinized social practices. In other words, because we divide physical space up, we can more easily routinize our behaviors. Thus, certain kinds of social practices occur in specific places and times. Regionalization varies by form, character, duration, and span.

The *form* of the region is given in terms of the kinds of barriers or boundaries that are used to section it off from other regions. The form allows greater or lesser possible levels of co-presence. When you stop and talk with someone in the hallway, there is a symbolic boundary around the two of you that is fairly permeable; it is very possible that others could join in. However, when you go into the men's or women's restroom, there is a physical and symbolic barrier that explicitly limits the possibility of co-presence.

The *character* of the region references the kind of social practices that can typically take place within a region. For example, people have lived in houses for centuries, but the character of the house has changed over time. In agrarian societies, the home was the center of the economy, government, and family; but in modern capitalist societies, the home is the exclusive domain of family and is thus private rather than public.

The *duration* and *span* of the region refer to the amount of geographic space and to the length or kind of time. Certain regions are usually available for social practices only during certain parts of the day or for specific lengths of time; the bedroom is an example in the sense that it is usually associated with "sleep time." Regions also span across space in varying degrees. So, a coliseum gives unique opportunities for co-presence and social activities when compared to an airplane.

We come now to the principal force behind Giddens's structuration and time–space distanciation. The word *ontology* refers to the study of existence. The point in using it here is that the way in which humans and their world exist is unique; as we've seen, it's meaningful. The reality of the human world is existentially moored in meaning, which is fallible, mutable, and uncertain. As the philosopher Ernst Cassirer (1944) puts it, "No longer can man confront reality immediately; he cannot see it, as it were, face to face. Physical reality seems to recede in proportion as man's symbolic activity advances. Instead of dealing with the things themselves man is in a sense constantly conversing with himself" (p. 42). According to Giddens, if people ever notice this about their reality, they will suffer deep psychological angst. We are motivated, then, as a result of this unconscious psychological insecurity about the socially created world, to make the world routine and thus taken-for-granted. Note that this anxiety is unconscious—it isn't usually experienced; but when it is, it is felt as a diffuse, general sense of unease.

According to Giddens (1990), **ontological security** refers to the feelings of "confidence that most humans [sic] beings have in the continuity of their self-identity and in the constancy of the surrounding social and material environments of action" (p. 92). Giddens argues that the fundamental trust of ontological security is generally produced in early childhood and maintained through adult routines. Because most of the social practices in our lives are carried out by routine, we experience trust in the world, due to its routine character, and we can take for granted the ontological status of the world. In pre-modern societies, trust and routine in traditional institutions covered up the contingency of the world. Kinship and community created bonds that reliably structured actions through time and space. Religion provided a cosmology that reliably ordered experience. And tradition itself structured social and natural events, because tradition by definition is routine.

In modern societies, however, none of these institutional settings produces a strong sense of trust and ontological security. According to Giddens, those needs are met differently: Routine is integrated into abstract systems, pure relationships substitute for the connectedness of community and kin, and reflexively constructed knowledge systems replace religious cosmologies—but not with the certainty or the psychological rewards of pre-modern institutions. The result is that ontological insecurity—anxiety regarding the "existential anchoring of reality" (Giddens, 1991, p. 38)—is a greater possibility in modern rather than traditional societies.

Concepts and Theory: The Contours of Modernity

We have now laid the groundwork for Giddens's understanding of how society works in general: Actors are motivated to routinize social actions and interactions by the psychological need for ontological security. These routines serve to stretch out face-to-face encounters through time and space as actors use different modalities to express the social structures of signification, domination, and legitimation through their social practices. This constant structuration produces different institutional orders that, along with regionalization, work to stabilize routine. Routinization and the institutional orders that it generates stabilize time–space distanciation and thus give the individual a continual basis of trust in her or his social environment, which, in turn, provides the individual with ontological security.

Thus, in Giddens's scheme, society isn't structured; that is, it doesn't exist as an obdurate object with an independent existence. The important point here is that society by its nature is continually susceptible to disruption or change. This constant possibility is, of course, what creates the diffuse and unconscious sense of insecurity that people have about the reality of society. However, this possibility is also what makes modernity an important issue, for both the process of structuration and for the person. In the next section, we will think about how living in modernity influences our experience of our self and others. But for now simply think about how the dynamic quality of modernity radically changes structuration and time–space distanciation. There are four analytically distinct factors that produce the dynamism of modernity: radical reflexivity, the separation of time and space, disembedding mechanisms, and globalization. As we'll see, though we can separate these areas analytically, they empirically reinforce one another.

Radical Reflexivity

Giddens sees reflexivity as a variable, rather than a static condition. He argues that modernity dramatically increases the level of reflexivity. Previous to this time, people didn't think much about society. In fact, the entire idea of society as an entity unto itself wasn't really conceived of until the work of people like Montesquieu and Durkheim. Today, we are quite aware of society and we think deliberatively about our nation and the organizations and institutions in which we participate.

Progress and reflexivity are intrinsically related. It only takes a moment's reflection to see that progress demands reflexivity. It is endemic in modernity because every social unit must constantly evaluate itself in terms of its mission, goals, and practices. However, the hope of progress never materializes—the ideal of progress means that we never truly arrive. Every step in our progressive march forward is examined in the hopes of improving what we have achieved. Progress becomes a motivating value and a discursive feature of modernity, rather than a goal that is ever reached.

Here's a real-life example. Chances are good that you are attending an accredited college or university. Schools of higher education are certified by regional accrediting organizations. Being accredited allows you as a student to qualify for federal financial aid and to transfer credits from one college to another, and it allows professors to apply for federal grants for research. At my university, we just finished with our reaccreditation self-study. This self-study took 2 years to complete. Even though 2 years seems like a long time, we actually began preparing for the self-study the 2 years previous by evaluating our mission statement in light of what we knew to be the new criteria of accreditation.

Out of the first study came a new mission statement that was then used during the following study to reevaluate every aspect of the university (notice the reflexive element). The self-study produced recommendations for the next 10 years, and the study and its recommendations were scrutinized by a committee of academics and administrators sent by our regional affiliation. Changes were and will be implemented as a result of the study. The interesting thing to me is that 80 to 90% of the study deals with things that are only tangentially related to actual learning, which is what we think the university is about. Most of the study addresses symbolic or political issues that have little to do with what happens in the classroom or in your learning experience. The greater proportion of the changes, then, would not have come about except for their symbolic or political values and reflexive organization.

Modern organizations are bureaucratic in nature and are thus bound up with rational goal setting, recursive practices, and continual reflexivity. For example, the reaccreditation study I just mentioned will be repeated in 10 years and every 10 years thereafter. This year my department is doing its self-study, and it gets repeated every 5 years. When I worked for Denny's restaurants as a manager, we had corporate plans that helped form the regional plans that helped create the unit plans, which strongly influenced my personal plans as a manager. Depending on the level, those plans were systematically evaluated every 1 to 5 years. Modern organizations, institutions, and society at large are thus defined through the continued use of reflexive evaluation.

One further point about radical reflexivity: It forms part of our basic understanding of knowledge and rational life. Modern knowledge is equivalent to scientific knowledge, and part of what makes knowledge scientific is continual scrutiny and systematic doubt. This understanding of knowledge is woven into the fabric of our culture. Every child in the United States receives training in what is called scientific literacy. According to the National Academy of Sciences (1995), "This nation has established as a goal that all students should achieve scientific literacy. The *National Science Education Standards* are designed to enable the nation to achieve that goal. They spell out a vision of science education that will make scientific literacy for all a reality in the 21st century." Thus, children in the United States are systematically trained to be reflexive about knowledge in general.

Emptying Time and Space

In this section, it is very important for you to keep in mind what Giddens means by time–space distanciation—it's his way of talking about how our behaviors and actions are patterned and are thus somewhat predictable. Therefore, whatever happens to time and space in modernity influences the patterns of interaction that make up society. With that in mind, Giddens argues that the *separation of time and space* is crucial to the dynamic quality of modernity.

In order to understand how time and space can be emptied, we have to begin by thinking about how humans have related to time and space for most of our existence. Up until the beginnings of modernity, time and space were closely linked to natural settings and cycles. People have always marked time, but it was originally associated with natural places and cycles. The cycle of the sun set the boundaries of the day, the cycle of the moon marked the month, and the year was noted by the cycles of the seasons. But the week, which is the primary tool we use to organize ourselves today, exists nowhere in nature. It's utterly abstract in terms of nature. Something similar may be said about the mechanical clock. Previous to the invention and widespread use of the mechanical clock, people regulated their behaviors around the moving of the sun. (See Roy, 2001, pp. 40–45; McCready, 2001.)

Thus, in modern societies, time and space have become abstract entities that have been emptied of any natural connections. Further, the concept of space itself has become stretched out and more symbolic than physical. As I mentioned, modernity is distinguished by the belief in progress. Progress implies change, and the emptying of time and space "serves to open up manifold possibilities of change by breaking free from the restraints of local habits and practices" (Giddens, 1990, p. 20). Making time and space abstract has also aided in another distinctive feature of modernity, the bureaucratic organization. Our lives are subject to rational organization precisely because time and space are emptied of natural and social relations. My students and I can all meet at 9:45 A.M. in the Graham building, Room 308, for class because time and space have been emptied. And Boeing manufacturing in California can order parts from a steel plant in China to be ready for assembly beginning in January because time and space are abstract.

The emptying of time and space means that time–space distanciation can be increased almost without limit, which is one of the defining characteristics of modernity. Traditional societies are defined by close-knit social networks that create high levels of morality and an emphasis on long-established social practices and relationships. Any social form that could break with the importance of tradition would have to be built upon something other than close-knit social groups. Modernity, then, is defined as the time during which greater and greater distances are placed between people and their social relations. As we'll see, increasing time–space distanciation and escalating reflexivity mutually reinforce one another, and together they create the dynamism of modernity—the tendency for continual change.

Institutions and Disembedding Mechanisms

In discussing the transition from traditional to modern society, many sociologists talk about structural differentiation, especially functionalists. The problem that Giddens sees in institutional differentiation is that it can't give a reasoned account of a central feature of modernity: radical time–space distanciation. However, thinking about institutions in terms of disembedding mechanisms does. Thus, Giddens claims that the distinction between traditional and modern institutions isn't differentiation so much as it is embedding versus disembedding. **Disembedding mechanisms** are those practices that lift out social relations and interactions from local contexts. Again, let's picture a kind of ideal type of traditional society where most social relationships and interactions take place in encounters that are firmly entrenched in local situations. People would live in places where they knew everybody and would depend upon people they knew for help. Distant situations, along with distant others, were kept truly distant. There are two principal mechanisms that lifted life out of its local context: symbolic tokens and expert systems.

Symbolic tokens are understood in terms of media of exchange that can be passed around without any regard for a specific person or group. There are a few of these kinds of tokens around, but the example *par excellence* is money. Money creates a universal value system wherein every commodity can be understood according to the same value system. Of necessity, this value system is abstract; that is, it has no intrinsic worth. In order for it to stand for everything, it must have no value in itself. The universal and abstract nature of money frees it from constraint and facilitates exchanges over long distances and time periods. Thus, by its very nature, money increases time–space distanciation. And, the greater the level of abstraction of money, such as through credit and soft currencies, the greater will be this effect.

The other disembedding mechanism that Giddens talks about is *expert systems.* Let's again think about a traditional community. If you lived in a traditional community and were going to have a baby, to whom would you go? If in the same group you experienced marital problems, where would you go for advice? If you wanted to know how to grow better crops or appease the gods or construct a building or do anything that required some form of social cooperation, where would you go? The answer to all these questions, and all the rest of the details of living life, would be your social network. If you wanted to grow better crops, you might go to your friend Paul whose fields always seem full and alive. For marital advice, you would probably go to your grandparents; for childbirth help, you'd go to the neighbor's wife who had been practicing midwifery for as long as you can remember.

Where do we go for these things today? We go to experts—people that we don't personally know who have been trained academically in abstract knowledge. But we don't really have to "go to" an expert to be dependent upon expert systems of knowledge. For instance, I have no idea how to construct a building that has many levels and can house a myriad of classrooms and offices, yet I'm dependent upon that expert knowledge every time I go to my office or teach in a classroom. Every time we turn on a computer or flick a light switch or start our car or go to buy food at the grocery store— in short, every time we do anything that is associated with living in modernity—we are dependent upon abstract, expert systems of knowledge. Systems of expert knowledge are disembedding because they shift the center of our life away from local contexts to dependence on abstract knowledge and distant others, who sometimes never appear.

Globalization

Giddens argues that four institutions in particular form the dynamic and time period of modernity: capitalism, industrialism, monopoly of violence, and surveillance. In terms of the dynamic of modernity, capitalism stands out. Capitalism is intrinsically expansive. It is driven by the perceived need for profit, which in turn drives the expansion of markets, technologies, and commodification.

Industrialization is of course linked to capitalism, but it has its own dynamics and relationships with the other institutional spheres. Industrialism, the monopoly of coercive power, and surveillance feed one another and create what is generally referred to as the industrial-military complex. A military complex is formed by a standing army and the parts of the economy that are oriented toward military production. Once a coercive force begins to use technology, it becomes not only dependent upon industrialism but also provides a constant impetus for more and better technologies of force and surveillance. A military complex by its very existence is not only available for protection, it is also in its best interests to instigate aggression whenever possible in order to expand its own base and the interests of its institutional partners.

From Chapter 9, you already know that capitalism naturally implies relationships among various nations, and if you take a moment, you'll see that the military

known and firm social and institutional relationships and expectations. This shift from the traditional, social self with clear institutional guidelines to the individual reflexive project was brought about because of the dynamics of modernity that we reviewed in the previous section.

The body is drawn into this reflexive project as well. Before radical modernity, the body was, for the most part, seen as either the medium through which work was performed or a vehicle for the soul. In either case, it was of little consequence and received little attention unless it became an obstacle to work or salvation. In radical modernity, on the other hand, the body becomes part of self-expression and helps to sustain "a coherent sense of self-identity" (Giddens, 1991, p. 99). The body becomes wrapped up with the reflexive project of the self in four possible ways: appearance, demeanor, sensuality, and through bodily regimes. We covered the first two ways in our chapter on Goffman, so will just take time to review the last two.

The body is involved in the self-project through *bodily regimes*. In radical modernity, "we become responsible for the design of our own bodies" (Giddens, 1991, p. 102). The body is no longer a simple reflection of one's work but can become a canvas for a self-portrait. Capitalism, mass media, advertising, fashion, and medical expert knowledge have produced an overabundance of information about how the body works and what kinds of behaviors result in what kinds of body images. We are called upon to constantly review the look and condition of our body and to make adjustments as necessary. The adjustments are carried out through various body regimes of diet, exercise, stress-reducing activities (yoga, meditation), vitamin therapies, skin cleansing and repair, hair treatments, and so forth.

With the *organization of sensuality*, Giddens has in mind the entire spectrum of sensual feeling of the body, but the idea is particularly salient for sexuality. Together, mass education, contraceptive technologies, decreasing family size, and women's political and workforce participation created the situation where "today, for the first time in human history, women claim equality with men" (Giddens, 1992, p. 1). Giddens links women's freedom with the creation of an "emotional order" that contains "an exploration of the potentialities of the 'pure relationship'" and "plastic sexuality" (pp. 1–2). The idea of *plastic sexuality* captures a kind of sexuality that came into existence as sex was separated from the demands of reproduction. Plastic sexuality is an explicit characteristic of modernity. For the first time in history, sexuality could become part of self-identity. We should also note that since sexuality is part of the reflexive project of the self, it is subject to reflexive scrutiny and intentional exploration.

Pure Relationships

To begin our discussion of pure relationships, let's think about friendship. Giddens points out that early Greeks didn't even have a word for friend in the way we use it today. The Greeks used the word *philos* to talk about those who were the most near and dear, but this term was used for people who were in or near to family.

And the Greek *philos* network was pretty well set by the person's status position; there was little in the way of friends as we think of them, as personal choices.

In languages that did have a word for friend, these friends were seen within the context of group survival. Friends were the in-group and others were the out-group. The distinction was between friend and enemy, or, at best, stranger. Keep in mind that groups were far more important then than they are now because individual survival was closely tied to group affiliations and resources. A friend was someone you turned to in time of need; thus the values associated with friendship were honor and sincerity. Today, however, because of disembedding mechanisms and increased time–space distanciation, not all friends are understood in terms of in-group membership and actual assistance. The individual can have distant friends and is enabled and expected to take care of her- or himself (the reflexive project).

A fundamental change has thus occurred in friendships: from friendship with honor based on group identity and survival to friendship with authenticity based on a mutual process of self-disclosure. Rather than trust being embedded in social networks and rituals, trust in modernity has to be won, and the means through which this is done is self-evident warmth and openness. By implication, this authenticity and self-regulation provide the personal, emotional component missing in trust in the abstract systems of modernity.

Intimate relations in modernity are thus characterized by *pure relationships*—friendships and intimate ties that are entered into simply for what the relationship can bring to each person. Remember that traditional relationships were first set by existing networks and institutions and the motivation behind them was usually social, not personal. For example, most marriages were motivated by politics or economics (not by love) and were arranged for the couple by those most responsible for the social issues in question (not by the couple themselves). This is the way in which modern relationships are pure: They occur purely for the sake of the relationship. Most of our relationships are not anchored in external conditions, like the politically or economically motivated marriage. Rather, they are "free-floating." The only structural condition for a friendship or marriage is proximity: We have to be near enough to make contact. But with modern transportation and communication technologies, our physical space is almost constantly in motion and can be quite far-ranging, and we have "virtual" space at our fingertips.

In addition to the free-floating and pure nature of these relations, they are also reflexively organized, based on commitment and mutual trust, and focus on intimacy and "self" growth. Like the reflexive project of the self, relationships are reflexively organized; that is, they are continually worked at by the individuals, who tend to consult an array of sources of information. The number of possible sources for telling us how to act and be in our friendships and sexual relations is almost endless. Daytime television is filled with programming that explores every facet of relationships; the magazine rack at the local supermarket is a cornucopia of surveys and advice on how to have the best (fill in the blank with any aspect of an intimate relationship); it's estimated that over 2,000 new self-help book titles are published every year in the United States; and the Internet resources available for improving relationships are innumerable. Most of us have taken a relationship quiz with our partner at some point (if you haven't, just wait, it's coming), and all of us have asked of someone the essential question for relationships that are reflexively organized: "Is everything all right?" This kind of communication is a moral obligation in pure relationships; the gamut of communication covers everything from the mundane (How was your day at work?) to the serious ("Do you want to break up with me?").

Choice and Life Politics

Along with the accelerating changes in modernity, there has been a shift from emancipatory politics to life politics. *Emancipatory politics* is concerned with liberating individuals and groups from the constraints that adversely affect their lives. In some ways, this type of political activity has been the theme of modernity—it was the hope that democratic nation states could bring equality and justice for all. And, in some respects, this theme of modernity has failed. We are more than ever painfully aware of how many groups are disenfranchised.

Life politics, by way of contrast, is the politics of choice and lifestyle. It is not based on group membership and characteristics, as is emancipatory politics; rather, it is based on personal lifestyle choices. We have come to think of choice as a freedom we have in the United States. But it is more than that—it has become an obligation. Choice is a fundamental element in contemporary living. This principality of choice is based on disembedding mechanisms and time–space distanciation, and results in, as we've seen, the reflexive project of the self. Part of that project comes to be centered on the politics of choice.

Mass media also plays a role in creating choice by facilitating mediated experiences. *Mediated experiences* are in contrast to social experiences that take place in face-to-face encounters and are created as people are exposed to multiple accounts of situations and others with whom they have no direct association through time and space. Every time you watch television or read a newspaper, you are exposed to lives to which you have absolutely no real connection. Like so many other features of modernity, this stretches out co-presence but it also creates a collage effect. The pictures and stories that we receive via the media do not reflect any essential or social elements. Instead, stories and images are juxtaposed that have nothing to do with one another. The picture we get of the world, then, is a collage of diverse lifestyles and cultures, not a direct representation.

As a result of being faced with this collage, what happens to us as individuals? One implication is that the plurality of lifestyles presented to us not only *allows* for choice—it *necessitates* choice. In other words, what becomes important is not the issue of group equality, but rather, the insistence on *personal* choice. What is at issue in this milieu is not so much political equality (as with emancipatory politics) as inner authenticity. In a world that is perceived as constantly changing and uprooted, it becomes important to be grounded in one's self. Life politics creates such grounding. It creates "a framework of basic trust by means of which the life span can be understood as a unity against the backdrop of shifting social events" (Giddens, 1991, p. 215). Life politics, then, helps to diminish the possibility and effects of ontological insecurity.

A good example of life politics is *veganism*—the practice of not eating any meat or meat byproducts. Not only is eating flesh avoided, but also any products with dairy, eggs, fur, leather, feathers, or any goods involving animal testing. One vegan I know summed it up nicely when she said, veganism "is an integral component of a cruelty-free lifestyle." It is a political statement against the exploitation of animals, and for some it is clearly a condemnation of capitalism—capitalism is particularly responsible for the unnatural mass production of animal flesh as well as commercial animal testing. Yet, for most vegans, it is a lifestyle, one that brings harmony between the outside world and inner beliefs, and not necessarily part of a collective movement.

However, it would be wrong to conclude that life politics are powerless because they do not result in a social movement. Quite the opposite is true. Life politics springs from and focuses attention on some of the very issues that modernity represses. What life politics does is to "place a question mark against the internally referential systems of modernity" (Giddens, 1991, p. 223). Life politics asks, "Seeing that these things are so, what manner of men and women ought we to be?" In traditional society, morality was provided by the institutions, especially religion. Modernity has wiped away the social base upon which this kind of morality was based. Life politics "remoralizes" social life and demands "renewed sensitivity to questions that the institutions of modernity systematically dissolve" (p. 224). Rather than asking for group participation, as does emancipatory politics, life politics asks for self-realization, a moral commitment to a specific way of living. Rather than being impotent in comparison to emancipatory politics, life politics "presage[s] future changes of a far-reaching sort: essentially, the development of forms of social order 'on the other side' of modernity itself" (p. 214).

Summary

- According to Giddens, the central issue for social theory is to explain how actions and interactions are patterned over time and space; or, to use Giddens's terms, social theory needs to explain how the limitations inherent within physical presence are transcended through time–space distanciation. There are two primary ways through which this occurs: the dynamics of structuration and routinization.
  o Structuration occurs when people use specific modalities to produce both structure (rules and resources of signification, domination, and legitimation) and practice (physical co-presence). Thus, the very method of structuration reflexively and repeatedly links structure and person and facilitates time–space distanciation.
  o Routinization is psychologically motivated by a diffuse need for ontological security. The reality of society is precarious because it depends on structuration, which is reflexive and recursive. In other words, the process of structuration doesn't reference anything other than itself and it depends on ceaseless interactional work. This precariousness is unconsciously

sensed by people, which, in turn, motivates them to routinize their actions and interactions and to link their routines to physical regions and institutional orders, which further add stability.

o   Routinization was unproblematically achieved in traditional societies. People rarely left their regions and the institutional orders were slow to change. Modernity, however, is characterized by dynamism and increasing time–space distanciation. Dynamism and time–space distanciation are both directly related to radical reflexivity, extreme separation of time and place, the disembedding work of modern institutions, and globalization. These factors are related to the proliferation of science and progress, bureaucratic management, the mechanical clock and universal calendar, communication and transportation technologies, symbolic tokens and expert systems of knowledge, the military complex, and world capitalism.

•   As a result of radical modernity, the individual is lifted out of the social networks and institutions that socially situated the self by acquiring certain identities, knowledge, and life course markers. The modern individual is given the reflexive project of the self that is only internally referential. As part of the reflexive project of the self, the individual involves her- or himself in strategic life planning using expert systems of knowledge and mediated experiences, all of which are permeated with pervasive doubt. The reflexive project of the self involves constant evaluation and reevaluation based on possible new information (ever revised by experts and available through mass media) and self-reflection (How am I doing? Should I be feeling this way?). The reflexive project includes lifestyle politics in which the individual must reflexively work her or his way through continuously presented and expanding arenas of social existence. Individuals, then, become hubs for social change as they reflexively order their life in response to a constantly changing political landscape.

Building Your Theory Toolbox

Learning More—Primary Sources

•   To learn more about Giddens's theory of structuration, you should read *The constitution of society,* University of California Press, 1986.
•   For Giddens's theory of modernity, I recommend *Modernity and self-identity: Self and society in the late modern age,* Stanford University Press, 1991.

Learning More—Secondary Sources

•   An excellent encounter with Giddens's theory (not just a review) is Stjepan Gabriel Mestrovic's *Anthony Giddens: The last modernist,* Routledge, 1998.

Check It Out

- *Fluid modernity:* For an approach that is neither modern nor postmodern, I recommend you read Zygmunt Bauman, especially *Liquid modernity,* Polity Press, 2000; and *Postmodernity and its discontents,* New York University Press, 1997.

Seeing the World

- After reading and understanding this chapter, you should be able to answer the following questions (remember to answer them *theoretically*):
  - o What is time–space distanciation and why is it the central question for Giddens? How does modernity affect time–space distanciation?
  - o What are social structures, in Giddens's scheme? How do they exist and what do they do?
  - o What are modalities of structuration? What are the three modalities? What is the function of modalities of structuration?
  - o What are the three institutional orders and how are they created?
  - o What are practical and discursive consciousnesses and how do they fit in with reflexive monitoring?
  - o What is the unconscious motivation in human interaction? What specific processes come about due to this motivation? How does each process vary? What are their effects?
  - o What are the main processes that produce the dynamic of modernity? There are at least four. Define each process and explain how it contributes to the dynamic character of modernity.
  - o What is the reflexive project of the self? How did it become individualized? How is the body involved and why do you think the body is important in this project?
  - o Explain the differences between emancipatory and lifestyle politics. Why is lifestyle politics more prevalent today than emancipatory?
  - o What are pure relationships?

Engaging the World

- Giddens is one of the architects and proponents of what is known as the "third way" in politics. Using your favorite Internet search engine, look up "third way." What is the third way and how is Giddens involved? How can you see it related to his theory?

Weaving the Threads

- Compare and contrast Giddens's theory of structuration with Bourdieu's constructivist-structuralism approach. Specifically, how are patterns of behavior replicated in the long run? How does each one overcome the object–subject dichotomy? Do you find one approach to be more persuasive? Why or why not?
- Compare and contrast Wallerstein's, Luhmann's, and Giddens's views of modernity. I recommend you start with their defining characteristics of

Power

It comes to this: dwarf-throwing contests,
dwarfs for centuries given away
as gifts, and the dwarf-jokes
at which we laugh in our big, proper bodies.
And people so fat they can't
scratch their toes, so fat
you have to cut away whole sides of their homes
to get them to the morgue.
Don't we snicker, even as the paramedics work?
And imagine the small political base
of a fat dwarf. Nothing to stop us
from slapping our knees, rolling on the floor.
Let's apologize to all of them, Roberta said
at the spirited dinner table. But by then
we could hardly contain ourselves.

—Stephen Dunn (1996, p. 61)*

P ower. It's an uneasy word, a word we don't like to acknowledge in proper company. Perhaps we may even shy away from it in *im*proper company, because to speak it is to make it crass. And it is certainly a word that social scientists are uncomfortable yet obsessed with. Social scientists understand that power makes the human world go round, but they have a devil of a time defining it or determining where it exists. One of the reasons it is hard to define is that it is present in every social situation.

Some theorists see power as an element of social structure—something attached to a position within the structure, like the power that comes with being the president of the United States. In this scheme, power is something that a person can possess and use. Other theorists have defined power as an element of exchange. Others have seen power more in terms of influence. This is a more general way in which to think of power, because many types of social relationships and people can exercise influence.

Michel Foucault defines power differently from most that have come before him. Foucault asks us to see power in knowledge. That in itself isn't unusual. Marx saw a connection between power and knowledge; he called it ideology and false consciousness. Weber also recognized that knowledge and power are connected. He specifically saw that knowledge could be used as power the more society became bureaucratized. Foucault takes this idea of power and knowledge much further than either Marx or Weber. For Foucault, power is hidden and treacherous. It is found in truth and discourse, and carried out in bodies, minds, and subjectivities.

The Essential Foucault

Biography

We should begin this brief biography by noting that Foucault would balk at the idea that we need to know anything about the author in order to understand his work. Further, Foucault would say that any history of the author is something that we use in order to validate a particular reading or interpretation. Having said that, Foucault was born on October 15, 1926, in Poitiers, France. Foucault studied at the École Normale Supérieure and the Institut de Psychologie in Paris. In 1960, returning to France from teaching posts in Sweden, Warsaw, and Hamburg, Foucault published *Madness and Civilization,* for which he received France's highest academic degree, *doctorat d'État.* In 1966, Foucault published *The Order of Things,* which became a best-selling book in France. In 1970, Foucault received a permanent appointment at the Collège de France (France's most prestigious school) as chair of History of Systems of Thought. In 1975, Foucault published *Discipline and Punishment* and took his first trip to California, which came to hold an important place in Foucault's life, especially San Francisco. In 1976, Foucault published the first volume of his last major work, *The History of Sexuality.* The two other volumes of this history, *The Use of Pleasure* and *The Care of the Self,* were published shortly before Foucault's death in 1984.

Passionate Curiosity

In Foucault's (1984/1990b) own words, "As for what motivated me. . . . It was curiosity—the only kind of curiosity, in any case, that is worth acting upon with a degree of obstinacy: not the curiosity that seeks to assimilate what it is proper for one to know, but that which enables one to get free of oneself. After all, what would be the value of the passion for knowledge if it resulted only in a certain amount of knowledgeableness and not, in one way or another and to the extent possible, in the knower's straying afield of himself?" (p. 8). In brief, Foucault was interested in how ideas and subjectivities come into existence and how they limit what is possible. But Foucault's search was not simply academic, though it was that. As the above quote tells us, Foucault sought to understand his own practices "in relationship of self with self and the forming of oneself as a subject" (p. 6).

Keys to Knowing

power, knowledge, order, games of truth, discourse, counter-history, archaeology and genealogy, episteme, historical rupture, subject objectification, panopticon, human disciplines, governmentality, microphysics of power, sexuality and subjectivity

Foucault's Perspective: Truth Games

Foucault is a complex thinker and writer. As a result, trying to summarize Foucault's theory can be a frustrating experience. In writing this chapter, I had a continuing sense of incompletion. The more I wrote, the more I felt that I was leaving out. I mention this because I know that what I'm presenting in this book is a pared down version of Foucault. Yet I believe that in focusing on a select few of Foucault's major points, I can convey some sense of what he was trying to accomplish.

Stated succinctly, Foucault is interested in how power is exercised through knowledge or "truth" and how truth is formed through practice (note that with Foucault, we can use knowledge and truth interchangeably). His interest in truth isn't abstract or philosophical. Rather, Foucault is interested in analyzing what he calls *truth games*. His use of "games" isn't meant to imply that what passes as truth in any historical time is somehow false or simply a construction of language. Foucault feels that these kinds of questions can only be answered, let alone asked, after historically specific assumptions are made. In other words, something can only be "false" once a specific truth is assumed; Foucault is involved in uncovering *how* truth is assumed. Specifically, Foucault's interest in truth concerns the game of truth: the rules, resources, and practices that go into making something true for humans.

The idea of practice is fairly broad and includes such things as institutional and organizational practices as well as those of academic disciplines—in these practices, truth is formed. The idea also refers to specific practices of the body and self—these are where power is exercised. Most of us use the word practice to talk about the behaviors we engage in to prepare for an event, like band practice for a show. But practice has another meaning as well. This meaning is clear when we talk about a medical practice. When you go to your physician, you see someone who is "practicing" medicine. In this sense, practice refers to choreographed acts that interact with bodies—sets of behaviors that together define a way of doing something. This is the kind of practice in which Foucault is interested.

Foucault's Method: Counter-History

Foucault uncovered truth games by constructing what he called counter-histories. When most of us think of history, we think of a factual telling of events from the past. We are aware, of course, that sometimes that telling can be politicized, which is one reason we have "Black History Month" here in the United States—we are trying to make up for having left people of color out of our telling of history. But most of us also think that the memory model is still intact; it's just getting a few tweaks. Foucault wants us to free history from the model of memory. He really doesn't say anything directly about whether any particular history is more or less true; that's not an issue for him. History in all its forms is part of and generated by discourse. Thus, Foucault's concern is how the *idea* of true history is used. What Foucault wants to produce for us is a *counter-history*—a history told from a different point of view from the progressive, linear, memory model.

The important questions then become, why is one path taken rather than another? Why is the present filled with one kind of discourse rather than others? And what has been the cost of taking this path rather than all the other potentialities? Thus, a counter-history identifies

> the accidents, the minute deviations—or conversely, the complete reversals—the errors, the false appraisals, and the faulty calculations that gave birth to those things that continue to exist and have value for us; it is to discover that truth or being does not lie at the root of what we know and what we are, but the exteriority of accidents. (Foucault, 1984, p. 81)

Foucault uses two terms to talk about his counter-history, archaeology and genealogy. Though the distinctions are sometimes unclear, *archaeology* seems to be oriented toward uncovering the relationships among social institutions, practices, and knowledge that come to produce a particular kind of discourse or structure of thought. *Genealogy* may be better suited to describe Foucault's (1984) work that is concerned with the actual inscription of discourse and power on the mind and body: "Genealogy, as an analysis of descent, is thus situated within the articulation of the body and history. Its task is to expose a body totally imprinted by history and the process of history's destruction of the body" (p. 83). We could say that archaeology is to text what genealogy is to the body. In both cases, there is an analogy to digging, searching, and uncovering the hidden history of order, thought, madness, sexuality, and so on. The hidden history isn't necessarily more accurate—it's simply a counter-story that is constructed more in an archaeological mode than an historical one.

Foucault's Critical Perspective

What is Foucault's point in constructing counter-histories? Part of what he wants to do is expose the contingencies of what we consider reality, but to what end? Many critical perspectives are based on assumptions of what would make a better society. In other words, there must be something to which the current situation is compared to demonstrate what it is lacking. But Foucault sees it otherwise. For him, *the critical perspective in itself is sufficient because it opens up possibilities.* In fact, Foucault would argue that a utopian scheme only attempts to replace one system of impoverishment with another. The point is to keep possibilities always open, to keep people critically examining their life and knowledge system so that they can perpetually be open to the possibility of something else.

According to Foucault's scheme, an important part of what creates knowledge, order, and discourse is the presence of "blank spaces." Foucault (1966/1994b) pictures knowledge as a kind of grid. The boxes in the grid are the actual linguistic categories, like mammal, flora, mineral, human, black, white, male, and female. We are familiar with those parts of the grid; they form part of our everyday language. However, there is actually a more important part of the grid: the part of the grid that creates the order—the blank spaces between the categories. "It is only in the blank spaces of this grid that order manifests itself in depth as though already there, waiting in silence for the moment of its expression" (p. xx). The true power of a discourse or knowledge system is in the spaces between the categories. As Eviatar Zerubavel (1993) notes, "separating one island of meaning from another entails the introduction of some mental void between them. . . . It is our perception of the void among these islands of meaning that makes them separate in our mind, and its magnitude reflects the degree of separateness we perceive among them" (pp. 21–22).

These spaces are revealed most clearly in transgression. As an illustration, let's think about a little boy of about 3 or 4 years of age. He is playful, playing with the toys he's been given and emulating the role models he sees on TV and among the neighborhood children. But one day his father comes home and finds him playing with dolls. His father grabs the doll away and tells his son firmly that boys do not play with dolls. In this instance, the category of gender was almost invisible until the young boy unwittingly attempted to cross over the boundary or space between the categories. The meaning and power of gender waited "in silence for the moment of its expression."

This idea of space is provocative. A more Durkheimian way of thinking about categories would conceptualize the space between them as a boundary or wall. Using the idea of boundary to think about the division between categories is fruitful: Walls separate and prevent passing. The young boy in our example certainly came up against a wall, and many of us have felt the walls of gender, race, or sexism. But the idea of walls makes the use of categories and knowledge seem objective, as if they somehow exist apart from us, and this is not what Foucault has in mind.

Notice that the boy in our example was unaware of the "wall" until his father showed it to him. From Foucault's position, the wall of gender was erected in the father's gendered practices. Foucault's idea of space helps us think about the practices of power. Space, in this sense, is empty until it is filled—seeing space between categories rather than a wall makes us wait to see what will go there and *how* it goes there. Space is undetermined. Something can be built in space but the space itself calls our attention to potential. Foucault's research, his critical archaeology, fills in that potential—he tells us how that space became historically constructed in one way rather than any of the other potential ways.

Foucault's counter-history actually creates a space of its own. On one side, Foucault's archaeology of modernity uncovers the fundamental codes of thought that establish for all of us the order that we will use in our world. On the other side, Foucault sets the sciences and philosophical interpretations that explain why such an order exists. Between these two domains is a space of possibilities, a space wherein a critical culture can develop that sufficiently frees itself "to discover that these orders are perhaps not the only possible ones or the best ones" (Foucault, 1966/1994b, p. xx).

In other words, through the archaeology of knowledge, Foucault wants to not only expose the codes of knowledge that undergird everything we do, feel, and think; he also wants to set loose the idea that things might not be as they are. He wants to free the possibility of thinking something different. That possibility of thought exists in the critical space between—but in this case the space isn't specified, as it is in already existing orders. Foucault doesn't necessarily have a place he is taking us; he doesn't really have a utopian vision of what knowledge and practice ought to be. His critique is aimed at freeing knowledge and creating possibility; it's aimed at creating an empty space that is undetermined.

Concepts and Theory: The Practices of Power

According to Foucault, **power** isn't something that a person possesses, but it is something that is part of every relationship. Foucault tells us that there are three types of domains or practices within relationships: communicative, objective, and power. Communication is directed toward producing meaning; objective practices are directed toward controlling and transforming things—science and economy are two good examples, and practices of power, which Foucault (1982) defines as "a set of actions upon other actions" (p. 220), are directed toward controlling the actions and subjectivities of people. Notice where Foucault locates power—*it's within the actions themselves,* not within the powerful person or the social structure. Foucault uses the double meaning of "conduct" to get at this insight: Conduct is a way of leading others (to conduct an orchestra, for example) and also a way of behaving (as in "Tommy conducted himself in a manner worthy of his position."). Thus, we conduct others through our conduct.

However, Foucault's intent is not to reduce power to the mundane, the simple organization of human behavior across time and place. Rather, Foucault's point is that power is exercised in a variety of ways, many of which we are unaware. Power, then, becomes insidious. Power acts in the normalcy of everyday life. It acts by imperceptible degrees, exerting gradual and hidden effects. In this way, the exercise of power entices us into a snare that feels of our own doing. But how is power exercised? Where does it exist and how are we enticed? Foucault argues that power is exercised through the epistemes (underlying order) and discourses found in what passes as knowledge. The potential and practice of power exists in these epistemes and discourses that set the limits of what is possible and impossible, which in turn are felt and expressed through a person's relationship with her- or himself, in subjectivities—the way we feel about and relate to our inner self—and the disposition of the body.

Epistemes and Order

Order is an interesting idea. We order our days and lives; we order our homes and offices; we order our files and our bank accounts; we order our yards and shopping centers; we order land and sea—in short, humans order everything. How do we order things? One of the ways is linguistically: "Indeed, things become meaningful only when placed in some category" (Zerubavel, 1993, p. 5). But a deeper and more fundamental question can be asked: How do we order the order of things? In other words, what scheme or system underlies and creates our categorical schemes? We may use categories to order the world around us, but where do the categories get their order?

To introduce us to this question, Foucault (1966/1994b) tells a delightful story of reading a book that contains a Chinese categorical system that divides animals into those "(a) belonging to the Emperor, (b) embalmed, (c) tame, (d) sucking pigs, (e) sirens, (f) fabulous, (g) stray dogs, (h) included in the present classification, (i) frenzied, (j) innumerable, (k) drawn with a very fine camelhair brush, (l) et cetera, (m) having just broken the water pitcher, (n) that from a long way off look like flies" (p. xv). The thing that struck Foucault about this system of categories was the limitation of his own thinking—"the stark impossibility of thinking *that*" (p. xv, emphasis original). In response, Foucault asks an important set of questions: What sets the boundaries of what is possible and impossible to think? Where do these boundaries originate? What is the price of these impossibilities—what is gained and what is lost?

Foucault argues that there is a fundamental code to culture, a code that orders language, perception, values, practices, and all that gives order to the world around us. He calls these fundamental codes epistemological fields or the *episteme* of knowledge in any age. **Episteme** refers to the mode of thought's existence, or the way in which thought organizes itself in any historical moment. An episteme is the necessary precondition of thought. It is what exists before thought and that which makes thought possible. This foundation of thought is not held consciously. It is undoubtedly this preconscious character of the episteme that makes thought believable and ideas seem true.

Further, rather than seeing thought and knowledge as results of historical, linear processes, Foucault argues that discontinuity marks changes in knowledge. Most of us think that the knowledge we hold accumulated over time, that we have thrown out the false knowledge and replaced it with true knowledge as we progressively learned how things work. This evolutionary view of knowledge actually comes from the culture of science. It is the way we want to see our knowledge, not necessarily the way it is. Foucault argues that knowledge doesn't progress linearly. Rather, what we know and how we know it is linked to historically specific patterns of behavior, institutional arrangements, and economic and social practices that set the rules and conditions of discourse and the limits of our possibilities. And that historical path is marked by rupture: discontinuities and sudden, radical changes.

Think about this: What is Foucault saying that hasn't been said before? Others, like Berger and Luckmann, have said that knowledge is socially constructed. The importance of what Foucault is saying is that this idea of rupture implies that knowledge and truth are purely functions of institutional arrangements and practices and not the result of any real quest for truth. Thus, what counts as truth in any age—our own included—comes about through historically unique practices and institutional configurations. This implies not only that knowledge is socially constructed, but also, and more importantly, that *knowledge is nothing more and nothing less than the exercise of power.* This pure power is put into effect through discourse and the taken-for-granted ordering of human life

Discourse

Discourse refers to languages and behaviors that are specific to a social issue, like the discourse of race. In simple terms, a **discourse** is a way of talking about things. If you want to discuss music with a group of musicians, for instance, there is an acceptable discourse or language that you would use. It would include such words as key, modes, transposing, and so on. This discourse would be different from the one you would use to talk about baseball. You wouldn't normally tell your baseball team to hit the field and "tune up," nor would you tell a violinist to "bunt."

Foucault's interest in the idea of discourse is a bit more significant. First, he wants us to see beneath the surface of the word choice between "bunt" and "tune up." Foucault is interested in the rules and practices that underlie the words and ideas that we use. Discourse sets the possibilities of thought and existence. There is an obvious link between language and thought: We think in language. So, a discourse, with its underlying rules and practices, gives us a language with which to think and talk. That's a commonsensical statement, and we might be tempted just to accept it at face value. But using discourse as the basis of thought sets the boundaries of what is possible and impossible for us to think, so it is more profound than it might first appear.

The second thing that discourse does is to determine the position a person or object must occupy *in order to become the subject of a statement.* "I'm a man" is such a statement. For me to be a man, I must meet the conditions of existence that are set down in the discourse of gender. I not only have to meet those conditions for you; I must meet them for me as well, because the discourse sets out the conditions of subjectivity, how we think and feel about our self. Subjects, and the accompanying inner thoughts and feelings, are specific conditions within the discourse. As we locate ourselves within a discourse, we become subject to the discourse and thus subjectively answer ourselves through the discourse.

This work of positioning that discourse accomplishes is one of its most powerful acts. Think about it this way: It is extremely difficult to talk to someone about anything without positioning yourself within a discourse. There are discourses surrounding sports, family, gender, race, class, self-improvement, medicine, mass media, cars, trucks, and on and on. Once you begin to converse using a discourse, you automatically occupy a position within it that tells you how to think, feel, and act.

The third thing that Foucault wants us to see about discourse is that it is used instead of coercive force to impose order on a social group. Critically speaking, social order is always a problem for the elite in any society. One way to subjugate a population is through physical coercion. However, as we saw in Chapter 11, the use of force is costly and produces contrary effects. Discourse is used instead of force and is thus characterized by a will to truth and a will to power. In other words, there is political intention behind truth and power. What passes as truth and how truth is validated are dependent upon the discourse. And discourse intrinsically contains a will to power.

Let me give you a dramatic example. The attacks of September 11, 2001, were perpetrated by men who are either "terrorists" or "freedom fighters," depending on the discourse that is used. Within these discourses are legitimations and methods of reasoning that create these two different social meanings. Further, the discourses create the subjective experience of all the different peoples involved. The substance of one discourse is captured by the title of the report generated by the United States government: *The National Commission on Terrorist Attacks Upon the United States.* Clearly, the discourse in the United States defines the perpetrators as *terrorists* and the subjective experience of those in the United States as being innocently *attacked.* The substance of the other discourse is revealed in the title and opening lines of a document confiscated by the police in Manchester, England, during a search of an Al Qaeda member's home. The title of the document is *Declaration of Jihad,* and the opening lines are addressed to "those champions who avowed the truth day and night" (*Al Qaeda Training Manual,* n.d.). One discourse creates the meaning of attack and terrorist; the other creates the meaning of holy war and champion.

Making an Object Out of a Subject

For Foucault, then, power is not so much a quality of social structures as it is the practices or techniques that become power as individuals are turned into subjects through discourse. Foucault intends us to see both meanings of the noun "subject": as someone to control, and as one's self-knowledge. Here Foucault's unique interest is quite clear—perhaps the most insidious form of power is that which is exercised by our self over how we think and feel; it is the power we exercise in the name of others over our self.

In an interesting analysis, Foucault uses the state to illustrate both meanings of subject. State rule is usually understood in terms of power over the masses. While this is a true characteristic of the state, Foucault argues that the modern state also exercises individualization techniques that exercise power over the subjectivity of the person. Foucault talks about this form of ruling as **governmentality:** "the government of the self by the self in its articulation with relations to others" (Foucault, 1989, as quoted in Davidson, 1994, p. 119). Governmentality was needed because of the shift from the power of the monarch to the power of the state.

Under a monarchy, the power of the queen or king was absolute and she or he required absolute obedience, but the scope of that control was fairly narrow. The nation-state "freed" people from the coercive control of the monarchy but at the same time broadened its scope of control. The nation-state is far more interested in controlling our behaviors today than monarchies were 300 years ago. In governmentality, the individual is enlisted by the state to exercise control over him- or herself. This is partly achieved through expert, professional knowledge that comes from medicine and the social and behavior sciences. The state supports such scientific research, and the findings are employed to extend control, particularly as the individual uses and consults medicine, psychology, and other sciences.

A fundamental part of Foucault's argument about the practices of power is the historical shift to *objectification*. Obviously, if power is intrinsic to human affairs of all kinds, then people have always exercised power. However, the practice of power became something different and more insidious due to historical changes that objectified the subject of power. I think we can get a picture of this shift in how power is exercised over the person by comparing the roots, primary meanings, and transitive verb forms of *object* and *subject*.

I've listed the differences between them in Table 13.1, for easy comparison. All the references come from Merriam-Webster (2002). Notice the attitudes that the Latin roots imply. Objects are things that can be thrown away, whereas subjects are things that are placed or thrown under. With this root meaning, subjects are controlled, but there is still a relationship between the subject and the one in charge— subjects are *under,* controlled but still present. Objects, on the other hand, are *thrown away;* there is no continuing relationship with whoever is doing the throwing. Notice also the first meaning of each word. Even though the definition of object is talking about something we perceive, like seeing a tree in the distance, *the object is regarded as in the way.* The first meaning of subject still carries with it the notion of connected but controlled.

We are all probably familiar with the transitive verb "to objectify." It means to make something an object that isn't an object, and it also means to exist apart from any internal relationship. Interestingly, most of us are probably not familiar with the transitive verb "to subjectify." As a case in point, my word processor just highlighted "subjectify" as a misspelled word, yet it is a real word that appears in exhaustive dictionaries. We just rarely use the word, nor do we think about things becoming subjectified—we assume that we subjectively relate to everything about ourselves. But, according to Foucault, that is not the case in modernity. Today, we relate to our self, our body, and our sexuality as objects.

Foucault's work is found in a series of books that provide a counter-history to some of the objectifying power practices found in Western societies. These books detail madness and rationality, abnormality and normality, medicine and the clinic,

within and dissects the patient. With the clinical gaze, "Western man could constitute himself in his own eyes as an object of science" (Foucault, 1963/1994a, p. 197).

Modern medicine is thus created through a gaze that makes the body an object, a thing to be dissected, either symbolically or actually, in order to find the disease within it. The culture of the clinical gaze helped to create a general disposition in Western society to see the person as an object. This disposition, along with the human sciences, made the practices of power much more effective and treacherous—objects that can be thrown away are much easier to control than subjects that demand continuing emotional and psychic connections.

Concepts and Theory: Power Over the Subject

Thus, bodily regimens of exercise and diet, self-understanding and regulation of feelings and behaviors, all stem from medicine and the human sciences, which Foucault tells us make up the panopticon of modernity. But Foucault is interested in something deeper than the control of the body—he wants to document how we as individuals exercise social power over the way we relate to our own selves. Nowhere is this more clearly seen than in Foucault's counter-history of sexuality. In order to understand Foucault's intent, we will now briefly review Greek and modern ideas of sexuality.

Greek Sexuality

Ancient Greece was the birthplace of democracy and Western philosophy. There was, in fact, a connection between democracy and philosophy. In Athens, in response to an upheaval by the masses against their tyrannical leader, a politician named Cleisthenes introduced a completely new organization of political institutions called democracy (the rule of common people). Through democratic elections, the elite incrementally lost their advantage in the assemblies and the common people ruled. Unfortunately, the masses were susceptible to impassioned speech and ended up making several decisions that conflicted with one another or entailed high costs. This series of crises created a desire in the elite for absolutes: What are the truths upon which all decisions and governance should be based? Truth obviously couldn't be found simply through rhetoric; they believed that there had to be some absolutes upon which decisions could be based.

Along with other factors, this impetus helped produce the Greek notion of the soul. For the Greek, the idea of the soul captured all that is meant by the inner person: her or his mind, emotions, ethics, beliefs, and so on. But in reading Plato, it's also clear that the soul was seen to be hierarchically constructed. Within the soul, the mind is preeminent and alone is immortal. The emotions and appetites, though part of the soul, are lesser and mortal. Thus, reason is godlike and education, especially philosophy, is important for proper discipline.

It is important that we see the emphasis here. The mind, emotions, and bodily appetites are viewed hierarchically, but they are all seen as part of the soul. In order to get a sense of the relationships within the soul, let's take a look at a conversation that Plato (1993) sets up between Socrates and a group of students. Socrates speaks first:

> "Do you think that it's a philosopher's business to concern himself with what people call pleasures—food and drink, for instance?"
>
> "Certainly not, Socrates," said Simmias.
>
> "What about those of sex?"
>
> "Not in the least." . . .
>
> "Then it is your opinion in general that a man of this kind is not preoccupied with the body, but keeps his attention directed as much as he can away from it and towards the soul?"
>
> "Yes, it is." . . .
>
> "Then when is it that the soul attains to truth? When it tries to investigate anything with the help of the body, it is obviously liable to be led astray."
>
> "Quite so."
>
> "Is it not in the course of reasoning, if at all, that the soul gets a clear view of reality?"
>
> "Yes." (pp. 117–118)

Notice how Socrates views sex: He puts it on the same level as eating and drinking. Sex isn't something set aside and special. It is simply seen as a bodily appetite, on a par with eating and drinking. And these aren't a direct concern for the philosopher—they are only of indirect concern. If the bodily appetites get in the way of the search for reality or truth, then they are of concern, but only then. The point is to keep the mind free. A person shouldn't be preoccupied with the body, because too much attention on the body and its appetites will take her or his attention away from the quest for truth. This bit of dialogue sets us up well for the way Foucault talks about sex in Greek society.

In Greek society, sexuality existed as *aphrodisia*. This Greek word is obviously where we get our term *aphrodisiac*, but it had a much broader meaning for the Greeks. Foucault notes that neither the Greeks nor the Romans had an idea of "sexuality" or "the flesh" as distinct objects. When we think of sex, sexuality, or the flesh, we usually have in mind a single set of behaviors or desires. The Greeks, while they had words for different kinds of sexual acts and relations, didn't have a single word or concept under which they could all fit. The closest to that kind of umbrella term is *aphrodisia*, which might be translated as "sensual pleasures" or "pleasures of love," and more accurately the works and acts of Aphrodite, the goddess of love.

These works of Aphrodite, perhaps like the works of any god or goddess, cannot be fully categorized. To do so would limit the god. This lack of a catalog or objective specification of sexuality is exactly Foucault's point. In modern, Western society, particularly as expressed through Christianity, there is a definite way to index those things that are sexual, or the "works of the flesh." This identifiability is extremely important for the Western mind because sex is a moral issue; it, above all other things, defines immoral practices. So, what counts and doesn't count as sexual is imperative for us, but wasn't for the Greeks.

The Greeks also employed the idea of *chrēsis aphrodisiōn* to sexuality: the phrase means "the use of pleasures." The Greeks' use of pleasure was guided by three strategies: need, timeliness, and status. The strategy of need once again highlights Socrates' approach to sexual practices. As we've seen, in Ancient Greece, the relationship to one's body was to be characterized by moderation; but every person's appetites and abilities to cope are different. Thus, the Greek strategy was for the individual to first know his need—to understand what the body wants, what its limits are, and how strong the mind is.

The second strategy is timeliness and simply refers to the idea that there are better and worse times to have sexual pleasures. There was a particularly good time in one's life, neither too young nor too old; a good time of the year; and good times during the day, usually connected with dietary habits. The issue of time "was one of the most important objectives, and one of the most delicate, in the art of making use of the pleasures" (Foucault, 1984/1990b, p. 57). The last strategy in the use of pleasures was status. The art of pleasure was adapted to the status of the person. The general rule was that the more an individual was in the public eye, the more he should "freely and deliberately" adapt rigorous standards regarding his use of pleasures.

Rather than seeing sexuality as moral, the Greeks saw it in terms of ascetics. *Ascetics* refers to one's attitude or relationship toward one's self, and for the Greek this was to be characterized through strength. The word ascetics comes from the Greek *askētikos,* which literally translated means exercise. The idea here is not simply something we do, as in exercising control; it also carries with it a picture of active training. Here we see the Greek link between masculinity and virility. The virile man in Greek society was someone who moderated his own appetites. He was the man who voluntarily wrestled with his body in order to discipline his mind. The picture we see is that of an athlete in training. For example, the athlete knows that eating chocolate or ice cream can be very pleasurable. But while in training, the athlete willingly forgoes those pleasures for what she or he sees as a higher good. The result of this training is *enkrateia,* the mastery of one's self. It's a position of internal strength rather than weakness.

Training is always associated with a goal; there is an end to be achieved or a contest to be won. In this case, the aim of the Greek attitude toward sexuality is a state of being, something that becomes true of the individual in his daily life. This is the *teleology,* or ultimate goal, of sexuality, the fourth structuring factor that defines a person's relationship to sex. The goal for the Greek was freedom. We can again see this idea in the conversation with Socrates. Truth and reality were things to be sought after. Too much emphasis on sex, just like eating and drinking, can get in the way of this search. As Socrates (Plato, 1993) said, "surely the soul can reason best when it is free of all distractions such as hearing or sight or pain or pleasure of any kind" (p. 118).

Western Modern Sexuality

The Western, modern view of sex is quite different from the Greek. It is, in fact, quite different from that which developed in the East. Where Eastern philosophy and religion developed a set of practices intended to guide sexual behavior to its highest and most spiritual expression and enjoyment (for instance, kama sutra), the West developed systems of external control and prohibitions. Of course, a great deal of the impetus toward this view of sex was provided by the Christian church.

Part of this movement came from Protestantism with its emphasis on individual righteousness and redemption. Rather than being worthy of God because of Church membership and sacraments, Protestantism singled the individual out and made her or his moral conduct an expression of salvation and faith. But an important part was also played by the Counter-Reformation, a reform movement in the Catholic Church.

Confession and penance are sacraments in the Catholic Church. They are one of the ways through which salvation is imparted to Christians. The Counter-Reformation increased the frequency of confession and guided it to specific kinds of self-examination, designed to root out the sins of the flesh down to the minutest detail:

> sex . . . [in all] its aspects, its correlations, and its effects must be pursued down to their slenderest ramification: a shadow in a daydream, an image too slowly dispelled, a badly exorcised complicity between the body's mechanics and the mind's complacency: everything had to be told. (Foucault, 1976/1990a, p. 19)

This was the beginning of the Western idea that sex is a deeply embedded power, one that is intrinsic to the "flesh" (*the* vehicle of sin par excellence, as compared to the Greek idea of bodily appetites), and one that must be eradicated through inward searching using an external moral code and through outward confession.

While these Christian doctrines would have influenced the general culture, they would have remained connected to the fate of Christianity alone had it not been for other secular changes and institutions beginning in the 18th century, most particularly in politics, economics, and medicine. With the rise of the nation-state and science, population became an economic and political issue. Previous societies had always been aware of the people gathered together in society's name, but conceiving of the people as the population is a significant change. *The idea of population transforms the people into an object that can be analyzed and controlled.*

In this transformation, science provided the tools and the nation-state the motivation and control mechanisms (ability to tax, standing armies, and so on). The population could be numbered and analyzed statistically, and those statistics became important for governance and economic pursuit. The population represented the labor force, one that needed to be trained and, more fundamentally, born. At the center of these economic and political issues was sexuality:

> It was necessary to analyze the birthrate, the age of marriage, the legitimate and illegitimate births, the precocity and frequency of sexual relations, the

As an icon of drag, Barbie illustrates what feminists and culture critics have been saying for some years. In no uncertain terms, Barbie demonstrates that femininity is a manufactured reality.

(Rogers, 2003, p. 95)

Have you ever thought of Barbie as "an icon of drag"? Whether you have or not, the above quote announces in no uncertain terms that there is something else going on. Barbie isn't simply Barbie. If we take the quote seriously, and I'm recommending we do, then what else isn't what it seems? What else is a G.I. Joe action figure? What else is a magazine picture of a body builder? What else is a teenager with an electric guitar strapped on so low it's at his knees? What else is a man with a big pickup truck or a woman with makeup? A bigger question is, how would we "read" these diverse cultural objects?

As we've seen, many of the central issues in contemporary social and sociological theory revolve around the ideas of meaning and language. Symbolic interactionism sees meaning as essential to human beings, yet pragmatically achieved through face-to-face interactions. Berger and Luckmann let us see that meaning is an anthropological necessity that constitutes our reality through the processes of objectification. Niklas Luhmann challenged us to see worldwide social systems with communication at the core. And Foucault illustrated the link between knowledge and power: Meanings are historically contingent, subject to disruption, and become oppressive power structures through bio-power. With Derrida, we come to the end of meaning, language, and certainty. Yet, in the same moment, Derrida will show us endless layers of meaning and textual languages and will introduce us to the possibility of Barbie as an icon of drag. Here's a thought-provoking quote from Derrida (1967/1997) to start us off: "However the topic is considered, the problem of language has never been simply one problem among others" (p. 6).

The Essential Derrida

Biography

Derrida was born July 15, 1930, in El-Biar, Algeria. In 1949, Derrida moved to France and began his studies at the École Normale Supérieur in Paris, where he studied under Michel Foucault and Louis Althusser. He eventually taught at the École Normale from 1965 to 1984. In 1967, Derrida published three books— *Speech and Phenomena; Of Grammatology;* and *Writing and Difference*—that established him as one of the most important philosophers of the 20th century. Beginning in the early 1970s, Derrida split his teaching time between Paris and the United States, teaching at Johns Hopkins, Yale, and the University of California at Irvine. In his career, he was awarded honorary doctorates from Cambridge, Columbia, the New School, the University of Essex, and Williams College.

Passionate Curiosity

Derrida had a passion for the hidden aspects of academic and social life. He saw an immense field of possibilities, only some of which are chosen. The question is, what is left out and why?

Keys to Knowing

structuralism, signified and signifier, poststructuralism, arche-writing, trace, text, *différance*, play, de-centering, deconstruction

---

Derrida's Perspective: Presence Through Absence

Jacques Derrida is a philosopher with a broad range of influences. The most notable of his philosophical considerations come from existentialism and phenomenology. Phenomenology is interested in finding the basic elements of phenomena as they present themselves (either in pure consciousness or the lifeworld). Existentialism seeks to determine the basic contours of existing as a self-aware being, in a specifically authentic existence. I go into some detail with phenomenology in

Chapter 2 and existentialism in Chapter 16, so I won't do so here. But I encourage you to review those specific sections.

Actually, the main reason to bring these ideas up here is to note that Derrida positions his philosophy against these influences. Both phenomenology and existentialism privilege a "presence." That is, with both approaches, there are ways of existing that are privileged over others: For phenomenology it's the essential phenomenon and for existentialism it's the authentic subject. Derrida's point is that in privileging something, the philosophies are blinded to the contingencies, complications, and suppression that are part of every element. More importantly, the belief in "presence" runs contrary to the intrinsic nature of language itself, which is a defining feature of humankind. As we'll see, Derrida argues that language can only create presence through absence.

I should also mention that Derrida was a student of Michel Foucault. Derrida (1967/1978), in fact, presented a paper in 1963 that both gave homage to and critiqued his teacher: "Having formerly had the good fortune to study under Michel Foucault, I retain the consciousness of an admiring and grateful disciple. . . . [Yet] the disciple must break the glass, or better the mirror, the reflection, his infinite speculation on the master. And start to speak" (pp. 31–32).

Derrida challenged Foucault's concept of reason and his interpretation of Descartes (Boyne, 1990, p. 55), the particulars of which have more relevance for philosophers than sociologists. Thus, we aren't going to make much of their disagreement here. Instead, what we will find is that Foucault and Derrida complement one another in a number of ways. Though the bases for their arguments are different, they end by relativizing our world and by giving us a new way to critique knowledge, language, and power.

In reviewing Derrida's thinking, we are going to combine his perspective and theory. The more philosophically based a thinker is, the more this makes sense because the perspective is the theory. Thus, we'll start off considering the main perspective that Derrida is arguing with: structuralism (at least it's the most important for our purposes of understanding social theory). We will then consider Derrida's ideas of arche-writing, trace, de-centering, and deconstruction and semiotics (which is how Barbie becomes an icon).

Concepts and Theory: Linguistic Structuralism

Jacques Derrida is generally considered to be the father of **poststructuralism.** In fact, Charles C. Lemert (1990) describes Derrida's speech in 1966 to the first major international conference on structuralism as the symbolic beginning of poststructuralism. While the seeds of structuralism were formed in the work of Ferdinand de Saussure, Claude Lévi-Strauss, and to some extent Émile Durkheim, structuralism didn't take off as an important intellectual movement until the mid-20th century. It is ironic, yet fitting, that the "rupture" or "event" that ended structuralism was announced by Derrida at the first major conference focusing on structuralism. As Lemert notes, "post-structuralism was born along with, and as part of, structuralism"

(p. 232). We will see what Lemert means as we consider Derrida's ideas, but first we need to understand structuralism.

Generally speaking, **structuralism** argues that there are deep structures that underlie and generate observable phenomena or events. This is a more radical statement than is usually made when we talk about structure in sociology (such as structural-functionalism). For many sociologists, social structures are seen as influencing our lives; they help account for the patterned nature of human action and interaction. But social structures are usually seen as one of several influences. While we can talk about the poles of the debate in terms of structure versus agency, most sociologists acknowledge that interactions, culture, and structure all influence our behavior.

On the other hand, structuralism sees the power of structure as absolute. These structures work below the level of consciousness, and they don't simply influence or even determine our behaviors; they generate, create, and produce them. Everything that we see, think, feel, and do are in reality events or manifestations of the structure. While this may sound depressing to some of us, for structuralists this idea represented a ray of hope. If structure does operate in this strong fashion, then social science could indeed function as a science: We could discover the underlying structure and its laws and could then explain, predict, and perhaps control all human behavior.

There has always been some discussion about where this structure might be found, but the place where many structuralists turned was language. Linguistic structuralism began with the work of Ferdinand de Saussure (1857–1913). Saussure's emphasis was on language as speech, a system of vocal signs. Writing for Saussure (1916/1986) is something different from speech, perhaps secondary or even derivative: "Language is a system of signs expressing ideas, and hence comparable to writing, the deaf-and-dumb alphabet, symbolic rites, forms of politeness, military signals, and so on. It is simply the most important of such systems" (p. 15).

Saussure saw speech as the authentic voice that gives presence to writing, which is artificial and at its best only represents speech. As such, writing is seen as having a number of secondary qualities that make it inferior to speech. Speech is immediate and intimately connected to the speaker, whereas written text is spatially and temporally separated from the author. For example, notice the dates associated with the above quote and the dates of Saussure's life. He died 3 years before the first publication of his book on general linguistics (Saussure's book is a compilation of lecture notes). His book was published after he died, and we are still reading it and talking about it almost 100 years later. Though, thanks to technology such as videotape, we can hear and see many of those who are no longer living, the displacement of space and time is fundamental to writing, by its nature.

Saussure argued that the very nature of the sign adds to the constancy, and thus the efficacy, of language. A linguistic sign is a "psychological entity" with two elements: *the signified,* the idea or sound image that the word calls up; and *the signifier,* the actual word. These two elements are meaningfully connected in that a change in one necessitates a change in the other.

Saussure (1916/1986) likens this relationship to a piece of paper: "One cannot cut the front without cutting the back at the same time" (p. 111). Further, the relationship between the two elements is arbitrary; that is, there is nothing inherent within the sound image (the psychological impression that a sound makes on the hearer) that links it to the concept—there is no real reason that a rock is called a "rock." The arbitrary nature of the relation places emphasis on the social origin of the structure: The relationship between signifier and signified is neither a function of innate mental categories nor ultimately a feature of physical objects. But this concept also puts The structure of language outside of immediate agency: Because the relationship is arbitrary, it has no "reasonable basis" for discussion and thus lacks the necessary foundation for rapid change.

Notice also that Saussure doesn't make any reference to an object. He isn't saying that language is referential, that it refers to anything outside of itself. Objects are ordered through speech, not the other way around. Speech isn't a reflection of the world—our language isn't put together by the natural order of the world; it's the linguistic system that gives order to an otherwise formless world. Think about the boundary between the United States and Canada. Where is it? It isn't in the natural world; it's in our language, in the way we talk about that geographic space. The same is true for "lawns," "farmland," "rivers," "rocks," and obviously for such social entities as gender and race. Speech, then, is the structure of the world and the structure of our lives, according to Saussure. The physical world becomes what it is to us through language, and the social world with all of its behaviors and encounters comes to exist through language as well.

Not only does language precede and order the physical world, it precedes and orders our thoughts and ideas as well. According to Saussure (1916/1986), we didn't come up with speech to reflect our ideas any more than we invented speech as a representation of the natural world. Ideas didn't exist ready-made before speech. In fact, apart from language, thought "is simply a vague, shapeless mass" (p. 110). Most of us seem to want to believe that language is simply a representation, that it expresses something real about ideas and objects. We want to believe such things because it gives our world and thoughts a firm anchor. If language reflects or expresses something, then there is more to reality and to us than just language. However, while we are more comfortable believing otherwise, it is extremely difficult if not impossible to experience the world or have thought apart from language, which is the thrust of Saussure's argument: The structure of language precedes and creates the events of our lives and minds.

Further, Saussure (1916/1986) argues that language is a system of interdependent terms "in which the value of any one element depends on the simultaneous coexistence of all the others" (p. 113). The most defining feature of a sign, then, is its opposition to other signs. Linguistic elements are given meaning through differences, or more properly, *différence*: "*In language itself, there are only differences. . . .* [The] language includes neither ideas nor sounds existing prior to the linguistic system, but only conceptual and phonetic differences arising out of that system" (p. 118, emphasis original). *Ultimately, then, signs themselves are not important, for it is the differences between the signs that create and limit meaning.*

There are two specific types of relationships between linguistic terms. Within a sentence, whether it is a written text or a conversation, combinations of elements are supported and given meaning by *linearity*. That is, the combinations of words that can appear together in a sentence are limited. These limited combinations define the meaning of any one word that stands within a combination through opposition to every other element that comes before or after it. Saussure (1916/1986) termed this relationship "syntagmatic" (p. 122).

The other specific relationship that a sign may have is more conceptual and lies outside the immediate sentence. These are *associative relations* and, because the concept behind the sign suggests other like concepts, they constitute relations of equivalence. Saussure offers the example of an architectural column. The column has a certain relationship with the rest of the building that it supports; this arrangement of physical elements in space illustrates the syntagmatic relation. But if the column is known to be of the Doric style, it might suggest a mental comparison with other styles, even though none of the other styles are present in physical space.

All this is less complicated than it might seem. Let's use the word "guitar" as an example. There are at least two elements in that sign: the word itself and its image. Though the relationship between the two sign elements is arbitrary, once established they seem to be the same thing. But what do we mean when we say "guitar"? Obviously, it points to the idea, but in terms of simply the sign itself, how do we establish its meaning?

We do this in part through difference: One of the things that makes a guitar a guitar is that it doesn't have certain attributes that the idea of a violin has. They are both stringed instruments, but what make them unique is their differences. Thus, a guitar is a stringed instrument that is not a violin (there are, of course, many other differences we could name). This defining function of difference is most clearly seen with words that have only one other companion word. These words are dichotomously defined. A good example is heterosexual. The meaning of heterosexual is defined by its opposite: homosexual. To be heterosexual is to not be homosexual.

Another way words achieve meaning is through associative relations. The presence of one word may imply another, like "electric guitar" implies "acoustic guitar" or perhaps "drums," though the other words aren't present. Words also derive their meaning by the way they are used in sentences. Thus, "it's a critical approach to understanding theory" has a different meaning from "it's an approach to understanding critical theory"—the words are basically the same, but the different order creates a different meaning.

There are four important ideas to take from this short discussion of Saussure. First, for Saussure, speech is primary because it is immediate. Second, a sign contains two elements: the actual sign and its sound image (concept). Third, the relationship between the sign and its sound image is arbitrary, with no referential relationship to the physical world. Fourth, the meaning of any sign is thus established by its structured relations with other signs, through *différence*.

Concepts and Theory: Poststructuralism

Arche-Writing

Derrida opposes much of what we've just seen from Saussure (that's why he is a poststructuralist). The first counterargument that Derrida presents is that writing, and thus text, logically occurs prior to and is more important than speech. He completely reverses Saussure's view. Derrida isn't saying that people wrote before they spoke—he isn't talking about the historical instance or "invention" of writing. Rather, Derrida sees writing as a kind of archetype of all language. In fact, Derrida calls this idea **arche-writing.** The word *arche* comes from Greek and literally means "the beginning." It implies something that has existed from the beginning, as a first principle or elementary substance. Derrida is arguing that all language is based upon fundamental arche-writing. Language, whether spoken or written, *inscribes itself,* which is the act of writing.

This is a heady but fairly simple idea, and we've actually already talked about it. As Saussure pointed out, the world and our thoughts apart from language are only shapeless and indistinct masses. But Derrida tells us that Saussure didn't fully understand his own idea. Language inscribes, it imposes or *writes* itself on our thought and upon the world. As Eviatar Zerubavel (1993) puts it, "We transform the natural world into a social one by carving out of it mental chunks we then treat as if they were discrete, totally detached from their surroundings" (p. 5).

Derrida (1967/1997) characterizes this carving out as the violence of difference and the space of possibility: "There is, as the space of its possibility, the violence of the arche-writing, the violence of difference, of classification, and of the system of appellations" (p. 110). Language is the act of making difference—it is the violent inscription of other upon nature and experience, and vocal speech does not adequately reflect this function of language. In contrast to Saussure, then, Derrida argues that speech, not writing, is derivative and secondary. Before speech could occur, there had to be a primitive and essential writing. Thus, because it is language and because language by its nature inscribes itself on its subject, even speech is fundamentally a form of writing.

In arche-writing, we obliterate the proper or natural and cast everything in the "play of difference" (p. 110). Saussure said this too, but he didn't fully appreciate the insight. Saussure talked about *différence;* Derrida coins the term **différance.** Remember that Saussure said that meaning is a function of sign *différence,* or the differences between signs. According to Saussure, signs are structured in hierarchical or categorical linguistic fields, just as ape is related to lemur under the general system of primate, which is itself related to the categories of mammal, animal, mineral, plant, and so on. Derrida points out that the difference is much more profound and basic than Saussure recognized.

Saussure saw the differences that created meaning within the linguistic system itself. On the other hand, Derrida is arguing that all language is fundamentally *based* on difference—or what he calls *difference.* It isn't something that only happens within speech—it is what all language, spoken or written, does by its nature. What this means, Derrida tells us, is that the characteristics that were thought to be secondary features of written text, the separation of time and space from the author, are in fact primary features of language in all its forms. All language is based on difference, on separation. This means that speech is dependent upon writing, specifically upon arche-writing, the idea and possibility of inscription: "What Saussure saw without seeing, knew without being *able* to take into account . . . is that a certain model of writing was necessary . . . as instrument and technique of representation of a system of language" (Derrida, 1967/1997, p. 43, emphasis original).

Before going on, I want to point out something that Derrida is doing. I'm sure you noticed how similar the words *différence* and *différance* are. Yet, if you look closely, you can see the distinction. However, when spoken in French, the language of both Saussure and Derrida, they sound almost exactly alike. The differences between the two words are hidden in speech but revealed in writing. The point that Derrida is making, of course, is that writing reveals things about language and itself that speech hides. Derrida's writing is filled with such hidden meanings that are based upon the play of words, so much so that some commentators argue that Derrida's theory is nothing but word play, which it hides more than explains. However, Derrida's word plays are intentional: Reading Derrida is an experience within the linguistic theory he is explicating.

Trace

Derrida uses the word *trace* to talk about the common elements of words. We should think about trace as a kind of remnant or residue, a mark each word gives to another. Merriam-Webster (2002) gives "engram" as a synonym for this idea of trace. An *engram* is a "protoplasmic change in neural tissue hypothesized to account for the persistence of memory." In other words, it is an etching on the brain tissue, a mark or writing that is the trace of an experience that is left on the brain. It's the trace or engram that links the brain and experience—it's the physical imprinting of memory. Engrams link the individual with her or his experiences and enable the person to remember. In the same way, Derrida is arguing that all words are made up of and connected by traces of other words. And just as engrams are the basic elements of memory, so too are traces basic to meaning.

Saussure posited a basic element of speech: the sound image. The *sound image* is not the physical or objective sound we hear. The sound image is the structure upon which the appearance of sound is based. This idea may appear a bit odd to us, but philosophers who work in metaphysics propose these kinds of behind-the-scenes realities all the time. Philosophers in this tradition look for those elements that are so essential and simple that they can be applied to all things that exist, whether physical or spiritual. In a like manner, Saussure argued that the basic metaphysical element behind the sound of speech is the sound image. It is the psychic image that speech calls up in our minds—the "psychological impression of a sound, as given to him by the evidence of his senses" (Saussure, 1916/1986, p. 66). Actual speech is based on these sound images; they are the fundamental elements that undergird actual speech and enable it to make an impact on our minds.

We've already seen that for Derrida, writing is essential to speech, so the most elementary principle must exist in writing, not speech. That basic element is the trace, the smallest unit of writing. As with the sound image, we should probably think of the trace as existing essentially and thus being out of sight. Of the trace, Derrida (1967/1997) says, "*no concept of metaphysics can describe it.* . . . it is *a fortiori* anterior to the distinction between regions of sensibility, anterior to sound as much as to light. . . . The graphic image is not seen*. . . . And, the difference in the body of the inscription is also invisible" (p. 65, emphasis original). In other words, the idea of trace is really hard, if not impossible, to truly explain or give clear examples of. But we can understand the implication of what Derrida is saying.

Remember that for Saussure, meaning is based upon differences (*différence*) within the structure of language. However, as we've seen, Derrida (1967/1997) argues that all signs are made up of traces of one another and thus refer to one another: "Without a trace retaining the other as other in the same, no difference would do its work and no meaning would appear" (p. 62). Thus, according to Derrida, difference comes out of sameness. Let's think about this idea in simple terms before we look at what Derrida intends.

As we've seen, all words are related to all other words; they are connected to one another through traces. Yet meaning requires difference. Picture an impossible situation: Let's pretend you consult a dictionary to find the meaning of two words, cat and dog. You look up "cat" and the definition goes on for page after page after page. It seems endless. Exasperated, you look up the meaning of "dog," because you think cats may be complex creatures but dogs are really simple and straightforward. But you're disappointed because the definition for dog is just as long as that for cat. In fact, the definitions are exactly the same and contain every definition and word in the English language.

What can we glean from this example? Just this: If dog and cat both mean the same thing and that meaning contains all possible meanings, then the words "dog" and "cat" are pretty senseless. They don't really mean anything because they mean *everything*, and they mean each other. So, dog and cat must be different in order for them to mean anything. Take this illustration to the extreme and have every word in the dictionary defined by every word in the dictionary. Language would be meaningless in such a case.

We thus need difference for meaning to exist. Saussure said that difference is in the structure of language; yet Derrida says that language is fundamentally built on sameness through traces. So, how can we have meaning? Derrida gives us two answers. The first is that the perception of meaning is an effect of an endless play of *deference*. In other words, any single sign has an infinite number of connections to other signs. In order to make one sign different from another, we defer or hold off all other possible signs, differences, and meanings. What this implies is that the production of meaning requires an indefinite deferral of signs in an endless play of differences. Meaning, then, isn't an effect of structure—something that would forever settle the sense of a word, symbol, or idea—nor is meaning an issue of pragmatic negotiations in interaction (Chapter 1); nor is it the socially constructed world of objective culture (Chapter 2); nor is it the underlying premise of social systems (Chapter 10). Rather, meaning comes out of endless play among textual differences and deferrals. It is never set, and it never truly means anything.

The use of "play" in this context is intentional. Because there is no actual basis for or structure of language, there is no "real" meaning. This brings us to Derrida's (1967/1997) second answer and to the root of his poststructuralism. The basic linkage or structure of language is trace. Traces don't exist in this world or any other world, and no more in time than in space, yet they "constitute the *texts*, the chains, and the systems of traces" (p. 65, emphasis original). In fact, to even talk about trace is already a trace (using language to talk about language), which is what leads Derrida to conclude that *"the trace is in fact the absolute origin of sense in general. Which amounts to saying once again that there is no absolute origin of sense in general. The trace is the différance* which opens appearance . . . and signification" (p. 65, emphasis original). Derrida thus leaves us right where the extremity of our dog and cat example left us: In the final analysis, language is meaningless because meaning is based on deference, not difference.

De-Centering

We need to take at least one more step before we can fully appreciate what Derrida is saying. Derrida argues that humans have always sought a center to the totality or structure of language. This center brought presence to language. The use of "presence" here functions as a technical term; it implies being or existence. In other words, humans have always sought a center to their linguistic schemes that would make language authentic, true, or real. Ideas such as essence, existence, substance, subject, transcendentality, consciousness, God, man, and so on have given language a reason for its existence, a firm foundation upon which to stand, and an "invariable presence" (Derrida, 1967/1978, pp. 279–280).

Derrida points out two things about the center. First, a center that moors language to some reality is by definition outside of language or the totality. The idea of the Christian God is a good example. God exists outside of time, space, and language, and because of that external existence, believers think that He gives reason for the universe in its totality. However, for Derrida, the external nature of the center reveals a contradiction. A center that is located outside the totality is by definition not in the totality and thus cannot be at the center of the totality. As Derrida (1967/1978) puts it, "The center is not the center" (p. 279). What we see in the idea of a center is a desire to master anxiety—anxiety about the human mode of existence. Since this is so, "the entire history of the concept of structure . . . must be thought of as a series of substitutions of center for center" (p. 279). Thus, logically, there is no center to language, no firm foundation upon which to stand.

The second issue that Derrida wants to bring to our attention is that a rupture has occurred in the history of the center. Writing is a profound process—it has the ability to write itself (this is one reason why Lemert said that poststructuralism was born along with structuralism). Through writing, we inscribe or write the world around us. That's what happens when we use language to understand and make meaning out of the physical world and our experiences. Through writing, we chisel meanings on the world and make it different from what it was. This is the *différance* of which Derrida speaks, and it is the violence that we perpetrate on the world. But that's not all. The process of inscribing not only writes the world, it can also write itself—"language bears within itself the necessity of its own critique" (p. 284).

Think of it this way: If a linguist wanted to critique language, from where would the elements of the critique come? To help us see the answer, let's think about another social system—the economy. When Marx critiqued capitalism, he used the processes and dynamics of capitalism to do so. Truly powerful critiques always originate from within. The same is true for language—and this is what we've seen with

said that—text, text, all is text. But what I'm pointing out here is that we aren't just talking about books, articles, newspapers, and the like. Keep in mind what writing is: It's an inscribing, so anything that has that kind of practice associated with it is a text. Derrida mentions film, dance, music, sculpting, athletics, and military and political activities. "All this to describe not only the system of notation secondarily connected with these activities but the essence and the content of these activities themselves" (Derrida, 1967/1997, p. 9). Everything we do or create is textual, including Barbie.

Deconstruction follows this idea and assumes that all analysis is textual. Rather than reading words, images, or activities as representing something outside the text, the best kind of analysis or interpretation is one that recognizes that meaning is achieved through a close reading of how words or other signs are related to each other within a text. Deconstruction looks for ways in which the explicit and implicit meanings of words, phrases, images, or rhetorical devices are at odds with or support one another, are hidden, are revealed, or are assimilated and used in the text.

Deconstruction is a sustained examination of texts for inconsistencies and hidden work. Every text is based upon the work of writing, and writing always suppresses alternative meanings. Remember, to say anything, a text must hold other possible meanings at bay. And, because signs contain traces and meaning is based on *différance,* then "other possible meanings" implies *all* possible meanings. Texts are built through word play, which implies that the work in text building comes from hiding. Sometimes the work of hiding is intentional and sometimes it's inadvertent, but writing is the work of hiding.

So, when you "read" something—like a book, a movie, an automobile, or someone's attire (anything!)—what you are picking up on is the hidden work of the text. *The task of deconstruction, then, is to reveal what's hidden.* But as much as possible, the reader must let the text point out its own contingencies. Sometimes these are found in footnotes or parenthetical statements, but just as often they are found in comparing one element of the text with another to reveal its own inconsistencies.

Inconsistencies may also be found in binary oppositions and ways in which one is privileged over the other (as with masculinity over femininity—or, the other way around, depending on the text). In deconstruction, these oppositions may be played with by reversing their order to point out the implicit hierarchy in the text (like writing "It was a male doctor" in place of "It was a female doctor").

Deconstruction is particularly salient for texts that make a claim to authority, such as religious or political writings. In the end, the goal of deconstruction is for the text to deconstruct itself: to reveal its own illusion of authority and determinacy, to move our attention away from our assumptions of meaning to intertextual play, and to repudiate the authority of the author in order to assert the authority of the reader.

I don't intend the above to be a definitive rendering of the term deconstruction. In fact, Derrida himself would find such an attempt to be presumptuous and misguided (can you reason out why?). I want simply to give you a taste of deconstruction and the implications of Derrida's poststructuralism. It's best to let Derrida (1997/2001) state the case himself:

If deconstruction is possible, this is because it mistrusts any sort of periodization and moves, or makes its gestures, lines and divisions move, not only within the corpus [the overall work] in general, but at times within a single sentence, or a microscopic element of a corpus. Deconstruction mistrusts proper names: it will not say "Heidegger in general" says thus or so; it will deal, in the micrology of the Heideggerian text, with different moments, different applications, concurrent logics, while trusting no generality and no configuration that is solid and given. It is a sort of great earthquake, a general tremor, which nothing can calm. I cannot treat a corpus, or a book, as a whole, and even the simple statement is subject to fission. (p. 9)

From beginning to end, Derrida's work is a deconstruction of Western philosophy. Derrida characterizes the positions that speech and the idea of center have occupied in linguistics and philosophy as "*logocentrism*." The root word, *logos*, comes from Greek and means speech, logic, and rationality. The word logos also has a long history in the Christian church. Jesus is referred to as "the Word [logos] of God," and we are told in the Bible that "the Word was made flesh and dwelt among us." Jesus in this way is seen as the complete expression of deity—an expression of true fidelity, a one-to-one correspondence between Jesus (logos) and God. I think this way of understanding logos helps us see Derrida's intent in using logocentrism. The search inherent within metaphysics and the assumptions fundamental to structural linguistics are comparable to the religious belief in the deity of Christ. Derrida shatters that belief and leaves us with nothing but text.

Semiotics

In addition to deconstruction, Derrida's idea that everything is text implies a *semiotic* (based on the study of signs) analysis. Thus, for example, images in television commercials may be seen as signs whose meaning is read through the manner in which they are placed next to one another, just like the syntagmatic meanings of words. Such images can have denotative and connotative meanings as well, and thus reference entire myths and discourses.

The concepts I just used come from Roland Barthes. Barthes uses semiotics to analyze different cultural texts. He explains that signs have both denotative and connotative functions. As we've seen, signs are primarily composed of an expression (signifier) in relation to a content (signified); together this forms the sign's denotative meaning. But a primary sign may also form part of a more complex expression. The original sign, its expression and content, can become the signifier of a secondary sign that is linked to a new content; this union forms the connotative meaning.

Connotation becomes itself a sign system and is increasingly important in modern societies: "The future probably belongs to a linguistics of connotation, for society continually develops, from the first system which human language supplies to it, second-order significant systems" (Barthes, 1964/1967, p. 10). This idea of Barthes's suggests endless layering of signification systems, each within a larger context.

A good illustration of this idea is the word *black*. The word specifically refers to a color and to a race of people with dark skin—those are two of its denotative meanings. But the word also connotes depression, evil, mourning, sadness, gloomy, and deserving of condemnation. Obviously, what often happens is that second-order meanings are applied unthinkingly to first-order denotations.

Douglas Hofstadter (1985) talks about this issue in terms of "default assumptions" and "slippery slopes." Second-order meanings are always implied in first-order ones, and the mind will default to them much as a computer defaults to its set programming. This creates slippery slopes of meaning, where language can be intrinsically racist or sexist ("man" to mean humanity and the male of the species) without explicit intent on the speaker's part.

Barthes extends this idea of connotation to include myths. Generally speaking, myths are metaphorical narratives that are meant to be interpreted on two levels: the surface level, and a deeper level usually referring to existential questions about humanity and the meaning of life (Nöth, 1995, p. 374). But Barthes tells us that myth-like narratives function on the everyday level as well. They can be found not only in religion, but also in movies, billboards, TV advertising, and so on.

For Barthes, myths are second-order connotative systems. In Barthes's formulation, myths becomes an instrument of cultural criticism: They function primarily through mass media as ideological realities that serve to both oppress and hide the oppression. Thus, systems of connotation can link ideological messages to more primary, denotative meanings, making them appear more natural.

Sociology

What can sociology glean from Derrida's work? He obviously points us to a textual understanding of society and the lives we live. This textual understanding is based on the assumption that everything that exists for humans is an instance of writing—nothing more and nothing less. Thus, we can read sports, billboards, movies, histories, medicine, and so on as texts. But rather than a defined and defining approach like most linguists propose, Derrida gives us a critical method of deconstructing the texts in a way that undercuts rather than affirms the assumed realities and ideologies in back of the text.

Beyond that, Lemert (1990) points us to four uses of poststructuralism. He says that in the social sciences, we typically solve the problems or questions we pose with "reference to ideas like 'empirical reality'" (p. 244). Poststructuralism shatters the idea of a center; our texts therefore can't legitimately make reference to an empirical reality because every reality for humans is a written (inscribed) one. Thus, what we have isn't an empirical reality; it is a textual reality. What then can we conclude? Lemert gives us four propositions for poststructuralist sociology:

1. that theory is an inherently discursive activity;

2. that the empirical reality in relation to which theoretical texts are discursive is without exception textual;

3. that empirical texts depend on this relationship to theoretical texts for their intellectual or scientific value; and

4. that in certain, if not all, cases a discursive interpretation yields more, not less, adequate understanding. (p. 244)

Lemert (1990) goes on to use these contours of poststructuralist sociology to read the Vietnam War: "Important as it is to recent American, and global, history the reality of Vietnam is far from certain" (p. 246). Lemert then shows how most people's understanding of Vietnam isn't tied to direct experience, but rather it arises out of a strange mixture of texts—like films such as *The Deer Hunter* and *Apocalypse Now,* the war memorial in Washington, various novels and public presentations, conversations with friends and relatives, and so on.

Further, Lemert (1990) shows, "the war itself was discursive, a global inscription in which the United States sought to mediate its own sense of the irreality of world history" (p. 247). Lemert then demonstrates how various texts created the possibility, guided the acts, and continually produced the idea of the Vietnam War. Lemert isn't arguing that nothing happened on the ground in Vietnam, but he is arguing that the reality of what happened was and is being created through various texts, rather than the war having any "empirical reality."

Lemert (1990) leaves us with a subtle but scathing criticism of empirically based sociology. I quote it at length:

What then are the prospects for a post-structuralist sociology? One answer might be found in the fact, reported by Russell Jacoby (1987), that between 1959 and 1969, the crucial years of the war, the three leading political science journals published 924 pieces of which exactly one concerned Vietnam. Sociology did not do much better. In the forty-six years between 1936 and 1982, the *American Sociological Review* published 2,559 articles, of which a scant five percent concerned political and social issues of any kind. This does not speak well for social science's grasp of reality.

Quite possibly a post-structuralist sociology would do better, however high the stakes. It would not be difficult to do as well. (pp. 251–252)

Summary

- Derrida argues that everything people do is a form of writing, an etching of difference or "meaning" upon an otherwise blank slate. This implies that all humans have is text, an implication that is strengthened by the idea of trace: All words and signs contain traces of one another. Language thus writes itself and refers only to itself. Further, any idea of language having a center is removed first by the case that all proposed centers exist outside the language system, which means that they can't be centers within language, and second by the historical shift away from seeing European culture as the center against which all other cultures must be judged. Without a center, language has no claim to reality or authority.

- According to Derrida, all we have then are texts; these texts can best be understood by analyzing them to the point where they deconstruct themselves. This is done by in-depth reading of the text alone, searching for the way words, rhetoric, and opposition are both hidden and used in the text. The point of deconstruction is to allow the text to reveal its own illusions of authority and specificity.

- Derrida's emphasis on text also formed a solid philosophical tradition for semiotics. Barthes asks us to see that signs have not only denotative but also connotative meanings, which are implied meanings that are produced through analogous relationships to other signs. Connotative meanings can actually be more fundamental to a society, in that they reference or cue societal myths. Lemert tells us that, in general, sociology would be well served to remember that theory is an inherently discursive activity, and that this recognition will generally yield more insights into society than other approaches (because society itself is discursively, textually formed).

Building Your Theory Toolbox

Learning More—Primary Sources

- As I mentioned in the chapter, reading Derrida is an experience in post-structuralism. So, be prepared and have fun! Remember, Derrida is not only telling us something, he is also playing with text. I recommend you start with a reader and then move on to fuller explanations:
  - *The Derrida reader: Writing performances,* edited by Julian Wolfreys, University of Nebraska Press,1998.
  - *Of grammatology,* Johns Hopkins Press, 1997.
  - *Writing and difference,* University of Chicago Press, 1978.

Learning More—Secondary Sources

- A good secondary source is
  - *Jacques Derrida,* by Nicholas Royle, Routledge, 2003.

Check It Out

- *Barbie as an icon of drag?* Read Mary Rogers's book, *Barbie Culture,* Sage, 1999, to find out more.
- *Derrida (motion picture):* http://www.derridathemovie.com/home.html
- *An interview with Derrida:* http://www.hydra.umn.edu/derrida/so.html (This one is particularly interesting because the interviewer asks the wrong questions, but they are questions I think you or I might ask.)
- *Poststructuralism: A very short introduction,* by Catherine Belsey, Oxford University Press, 2002.
- *Society as text: Essays on rhetoric, reasons, and reality,* by Richard Harvey Brown, University of Chicago Press, 1987.

- *Reading media:* The idea that mass media can be read as text has produced a very extensive literature. One of the best places to begin exploring this idea is *Gender, race, and class in media: A text-reader,* edited by Gail Dines and Jean M. Humez, Thousand Oaks, CA: Sage.

Seeing the World

- After reading and understanding this chapter, you should be able to answer the following questions (remember to answer them *theoretically*):
  o Explain French structuralism (Saussure's theory), paying particular attention to the implications of signified and signifier and the structural basis of meaning.
  o Explain arche-writing, trace, *différance,* de-centering, and how they form the basis of poststructuralism. You specifically will need to explain how each of these issues contributes to the loss of certainty in meaning and representation.
  o What is deconstruction? How does deconstruction fit in with Derrida's theory? Give an example of how a "text" might be deconstructed. In other words, how can deconstruction be used to understand our social world?

Engaging the World

- Reread the portion on deconstruction, especially the long quote from Derrida. Now, deconstruct this chapter.
- What are some areas in which you think the method of deconstruction might yield important results? Find a text in one of those areas and deconstruct it. Remember to use issues that are internal to the text.

Weaving the Threads

- In this book, we've seen that phenomenology has informed the work of Berger and Luckmann (Chapter 2), Harold Garfinkel (Chapter 3), and now Derrida. How has this philosophy influenced these different theorists? What are the essential elements of a phenomenological approach? What do we gain by using it? What do we lose?
- Derrida and Foucault are sometimes both considered poststructuralists. Use the central concepts from Derrida and Foucault to form a more robust understanding of what poststructuralism is.
- Use the index of this book to look up the different definitions and explanations of meaning. Evaluate each of these approaches and create what you feel to be a clear, robust, and correct definition/explanation of meaning. Justify your theory.

CHAPTER 15

The End of Everything

Jean Baudrillard (1929–)

Photo: Copyright: Res Stolkiner. European Graduate School EGS, Saas-Fee, Switzerland. 2002.

Have you ever been to Disneyland or Disney World? Many of us have. And if you're like most of us, you were dazzled by spectacle, rode amazing rides, enjoyed the famous Disney characters (like Mickey, Donald, Snow White, and the rest), watched shows, ate junk food, and generally just lost yourself in a world that let you be a kid again. It's the Magic Kingdom, after all! But what if I told you that there's something wrong about Disneyland? "Wrong" may not be the right word . . . but what if I told you that there is more to Disneyland than meets the eye? Maybe it really is magic. Maybe Disney has supernatural powers! Maybe Disneyland not only transforms the people who visit it, it also transforms the entire Los Angeles area! Well, I would not have thought so either, but that's exactly what Jean Baudrillard thinks of Disneyland.

The way I'm talking about Disneyland sounds similar to what I said about Barbie in the chapter on Derrida. In some ways it probably is, but in other ways it's very different. With Barbie, the point I wanted to bring out was that the toy could be read in a field of signs. And Baudrillard probably wouldn't disagree. But there's a way in which reading Barbie as an "icon of drag" is meaningful. Barbie in a cultural text with other signs is a specific way of talking about Barbie. If you understand the terms I'm using theoretically, then you'll get my meaning. Baudrillard, however, might be saying something more profound or disturbing. With Derrida, all we have is text. But text does refer to something: other texts. Granted, that can be pretty disturbing, but with Baudrillard, *there is no reference at all:*

The systems of reference for production, signification, the affect, substance and history, all this equivalence to a "real" content, loading the sign with the burden of "utility," with gravity—its form of representative equivalence—all this is over with. (Baudrillard, 1976/1993b, p. 6)

It's all gone—all meaning, all reference. It's over with. Any equivalence to a reality is just an illusion. You see, Baudrillard is a postmodernist, and postmodernism is . . . Oh-oh, we just ran into trouble. The trouble is that postmodernism argues that there is no reference. The point I'm making is that the word "postmodernism" references something definable. However, because culture is fragmented and free-floating for postmodernists, "postmodernism" can't reference any single, cohesive idea. It doesn't exist as anything. This state of affairs generally leaves authors in a pickle: How can we say anything about something if it doesn't exist as a thing? And that dilemma is part of the frustration and the fun that we can have with the idea of postmodernism. If we accept that words don't have any specific cohesive references, then we either become dark in our writings and thoughts or we can be playful, intentionally using words and references that have multiple and perhaps contradictory meanings.

The Essential Baudrillard

Biography

Jean Baudrillard was born in Reims, France, on July 29, 1929. Baudrillard studied German at the Sorbonne University, Paris, and was professor of German for 8 years. During that time, he also worked as a translator and began his studies in sociology and philosophy. He completed his dissertation in sociology under Henri Lefebvre, a noted Marxist-humanist. Baudrillard began teaching sociology in 1966, eventually moving to the Université de Paris-X Nanterre as professor of sociology. From 1986 to 1990, Baudrillard served as the director of science for the Institut de Recherche et d'Information Socio-Économique at the Université de Paris-IX Dauphine. Since 2001, Baudrillard has been professor of the philosophy of culture and media criticism at the European Graduate School in Saas-Fee, Switzerland. Baudrillard is the author of several international best-sellers, the most controversial of which is probably *The Gulf War Did Not Take Place* (1995).

Passionate Curiosity

Baudrillard is essentially concerned with the relationship between reality and appearance; this is an issue that has plagued philosophers and theorists for eons. Baudrillard's unique contributions to this problem concern the effects of mass media and advertising. Baudrillard, then, is deeply curious about the effects

Marx's argument is that human labor creates exchange-value and is the source of capitalist profit.

Baudrillard counters by arguing that Marx is continuing to focus on materialist, and thus capitalist, ideas. Like species being, use-value is completely bound up with the idea of products. Use-value is oriented to a materialist world alone. It is filled with practical use that is used up in consumption. It has no value or meaning other than material. And, like species being, the idea of use-value validates the basic tenets of capitalism—the truth of human life is rooted in economic production and consumption. Exchange-value is materialist as well, because exchange-value is based on human, economic production. The idea of exchange-value also legitimates and substantiates instrumental rationality, the utilitarian calculations of costs and benefits. Again, rather than critiquing capitalism, Baudrillard sees Marx as legitimating it.

> The Marxist seeks a *good use* of economy. Marxism is therefore only a limited petit bourgeois critique, one more step in the banalization of life toward the "good use" of the social! . . . Marxism is only the disenchanted horizon of capital—all that precedes or follows it is more radical than it is. (Baudrillard, 1987, p. 60, emphasis original)

In place of use- and exchange-values, Baudrillard proposes the idea of *sign-value*. Commodities are no longer purchased for their use-value, and exchange-value is no longer simply a reflection of human labor. Each of these capitalist Marxian values has been trumped by signification. In postmodern societies, commodities are now purchased and used more for their sign-value than for anything else. As you can see, postmodernity is part of the linguistic turn in sociology.

In order to be able to compare these differences, I've laid them out in Table 15.1. I'll explain the historical transitions in a moment, but symbolic exchange was preeminent in traditional, non-capitalistic societies. Baudrillard will refer to this time period as the first and second phases of the sign. Use-value and exchange-value were dominant during the early stages of capitalism and began what Baudrillard calls the consumer society (the third stage of the sign). Sign-value corresponds to postmodernity, the fourth stage of the sign. I'd like you to notice the shifts in "meanings." Symbolic exchange has personal meaning: Its meaning is strong and clear for those participating. Under capitalism, meaning shifts to instrumental and market concerns—what will this object do for me, and what can I gain from it? In postmodernity, "meaning" shifts to signs in relation to other signs. This latter idea is Derrida's point exactly; Baudrillard is simply seeing it as part of an historical progression. But Baudrillard also takes Derrida's notion a step further.

Commodity Fetish

Let's look at one more idea from Marx before moving to Baudrillard's theory. Marx was very critical of the process of commodification. A commodity is simply something that is sold in order to make a profit. *Commodification as a process* refers to the way more and more objects and experiences in the human world are turned

Table 15.1   Value Chart

	Symbolic Exchange	Use-Value	Exchange-Value	Sign-Value
Essence	Gift	Function	Economy	Consumption
Value	Symbol	Instrument	Commodity	Sign
Meaning	Relation to subject	What object does	Market worth	Relation to other signs
Examples	Wedding ring	Automobile as transportation	The book costs $19.95	Clothing as fashion

into products for profit. To see what Marx means, simply compare your life with that of your great-grandparents. Chances are that many more of the objects and experiences in your life are purchased rather than made; they exist as commodities in your world. Commodification is intrinsic to capitalism.

Increasing commodification leads to *commodity fetish*. Marx used the term *fetish* in its pre-Freudian sense of idol worship. The idea here is that the worshipper's eyes are blinded to the falsity of the idol. Marx's provocative term has two implications that are related to one another. First, in commodity fetish, people misrecognize what is truly present within a commodity. By this Marx meant that commodities and commodification are based on the exploitation of human labor, but most of us fail to see it. Second, in commodity fetish there is a substitution. For Marx, the basic relationship between humans is that of production. But in commodity fetish, the market relations of commodity exchange are substituted for the productive or material relations of persons. The result is that rather than being linked in a community of producers, human relationships are seen through commodities, either as buyers and sellers or as a group of like consumers. Commodification and its fetish are among the primary bases of alienation, which, according to Marx, separate people from their own human nature as creative producers and from one another as social beings.

Again, Baudrillard argues that Marx's concern was misplaced and actually motivated by the capitalist economy. Marx's entire notion of the fetish is locked up with species being and material production. With commodity fetish, we don't recognize the suppressive labor relations that underlie the product and its value, and we substitute an alienated commodity for what should be a product based in our own species being. In focusing exclusively on materialism, "Marxism eliminates any real chance it has of analyzing the *actual process of ideological labor*" (Baudrillard, 1972/1981, p. 89, emphasis original). According to Baudrillard, ideology isn't based in or related to material relations of production, as Marx argued. Rather, ideology and fetishism are both based in a "*passion for the code*" (p. 92, emphasis original).

One way to understand what Baudrillard is talking about is to go back to Table 15.1. Human nature is symbolic and oriented toward meaning. In symbolic exchange, real meaning and social relationships are present. However, capitalism and changes in media have pushed aside symbolic exchange and in postmodernity have substituted sign-value. And, as we've seen with Derrida, sign-value is based on textual references to other signs, nothing else. The fetish, then, is the human infatuation with consuming sign-vehicles that are devoid of all meaning and reality. Thus, ideology "appears as a sort of cultural surf frothing on the beachhead of the economy" (Baudrillard, 1972/1981, p. 144). Signs keep proliferating without producing substance. This simulation of meaning is what constitutes ideology and fetish for Baudrillard. This implies that continuing to use the materialist Marxian ideas of fetish and ideology actually contributes to capitalist ideology, because it displaces analysis from the issues of signification.

Concepts and Theory: Mediating the World

Pre-Capitalist Society and Symbolic Exchange

Baudrillard's (1981/1994) theory is based on a fundamental assumption: "Representation stems from the principle of the equivalence of the sign and of the real" (p. 6). What Baudrillard means is that it is possible for signs to represent reality, especially social reality. Think about it this way: What is the purpose of culture? In traditional social groups, culture was created and used in the same social context. For the sake of conversation, let's call this "grounded culture," the kind of culture that symbolic exchanged is based on. In a society such as the United States, however, much of our culture is created or modified by capitalists, advertising agencies, and mass media. We'll call this "commodified culture."

Members in traditional social groups were surrounded by grounded culture; members in postmodern social groups are surrounded by commodified culture. There are vast differences in the reasons why grounded versus commodified culture is created. Grounded culture emerges out of face-to-face interaction and is intended to create meaning, moral boundaries, norms, values, beliefs, and so forth. Commodified culture is produced according to capitalist and mass media considerations and is intended to seduce the viewer to buy products. With grounded culture, people are moral actors; with commodified culture, people are consumers. Postmodernists argue that there are some pretty dramatic consequences, such as cultural fragmentation and unstable identities (for a concise statement, see Allan & Turner, 2000).

Beginning with this idea of grounded or representational culture, Baudrillard posits four phases of the sign. The first stage occurred in pre-modern societies. The important factor here is that language in pre-modern societies was not mediated. There were little or no written texts and all communication took place in real social situations in face-to-face encounters. In this first phase, the sign represented reality in a profound way. There was a strong correlation or relationship between the sign and the reality it signified, and the contexts wherein specific signs could be used were clear. In this stage, all communicative acts—including speech, gift giving, rituals, exchanges, and so on—were directly related to and expressive of social reality. Here is the birthplace of symbolic exchange.

The second phase of the sign marked a movement away from these direct kinds of symbolic relationships. It gained dominance, roughly speaking, during the time between the European Renaissance and the Industrial Revolution. While media such as written language began previous to the Renaissance, it was during this period that a specific way of understanding, relating to, and representing the world became organized. Direct representation was still present, but certain human ideals became to make inroads. Art, for example, was based on observation of the visible world and yet contained the values of mathematical balance and perspective. Nowhere is this desire for mathematical balance seen more clearly than in Leonardo da Vinci's work, *Proportions of the Human Figure*. Thinking of some of da Vinci's other paintings, such as the *Mona Lisa* and the *The Last Supper*, we can also see that symbols were used to convey mystery and intrigue.

The Dawn of Capitalism and the Death of Meaning

The third phase of the sign began with the Industrial Revolution. This is the period of time generally thought of as modernity. The Industrial Age brought with it a proliferation of consumer goods never before seen in the history of humanity. It also increased leisure time and produced significant amounts of discretionary funds for more people than ever before. These kinds of changes dramatically altered the way produced goods were seen. Here is where we begin to see the widespread use of goods as symbols of status and power. Thorstein Veblen (1899) termed this phenomenon *conspicuous consumption.*

Baudrillard (1970/1998) characterizes this era as the beginning of the **consumer society.** Here is where commodities begin their trek toward sign-value, and symbolic exchange becomes buried under an avalanche of commodified signs. In the consumer society, social relations are read through a system of commodified signs rather than symbolic exchange. Commodities become the *sign-vehicles* in modernity that carry identity and meaning (or its lack). For example, in modern society the automobile is a portable, personal status symbol. Driving an SUV means something different from driving a Volkswagen Beetle, which conveys something different from driving a hybrid. As this system becomes more important and elaborated, a new kind of labor eventually supplants physical labor, *the labor of consumption.* This doesn't mean the work involved in finding the best deal. The labor of consumption is the work a person does to place her-or himself within an identity that is established and understood in a matrix of commodified signs.

Baurdrillard's point is that most if not all of shopping done in the consumer society is driven by sign-value. Every commodity sits in a signed relationship with other commodities—from garden rocks for sale with the word "imagine" etched into them, to tractors, to clothing, to cheese, to the way you listen to music (on records, CDs, MP3 players), to where you shop, and even to what you carry your purchases home in (plastic, paper, or canvas reusable bags). The principal form of labor, then, in the consumer society is the work expended in being able to recognize the "correct" sign-values in the ever-changing schemes of commodified signs.

The fourth stage of the sign began shortly after WWII and continues through today. This fourth stage occurs in postindustrial societies. As such, there has been a shift away from manufacturing and toward information-based technologies. In addition, and perhaps more importantly for Baudrillard, these societies are marked by continual advances and an increasing presence of communication technologies and mass media. Mediated images and information, coupled with unbridled commodification and advertising, are the key influences in this postmodernity. The cultural logic has shifted from the logic of symbolic exchange in pre-capitalist societies, to the logic of production and consumption in capitalist societies, and finally now to the logic of simulation. For Baudrillard (1976/1993b), postmodernity marks the end of everything.

> The end of labor. The end of production. . . . The end of the signifier/signified dialectic which facilitates the accumulation of knowledge and of meaning. . . . And at the same time, the end simultaneously of the exchange value/use value dialectic which is the only thing that makes accumulation and social production possible. . . . The end of the classical era of the sign. (p. 8)

Baudrillard (1981/1994) posits that the postmodern sign has "no relation to any reality whatsoever: it is its own pure simulacrum" (p. 6). These kinds of signs are set free from any constraint of representation and become a "play of signifiers . . . in which the code no longer refers back to any subjective or objective 'reality,' but to its own logic" (Baudrillard, 1973/1975, p. 127). Thus, in postmodernity, a fundamental break has occurred between signs and reality. Signs reference nothing other than themselves; they are their own reality and the only reality to which humans refer. These seem like brash and bold claims, but let's look at Baudrillard's argument behind them.

Concepts and Theory: Losing the World

Entropy and Advertising

First, Baudrillard argues that there is something intrinsic in transferring information that breaks it down. This is an important point: Anytime we relate or convey information to another, there is a breakdown. So fundamental is this fact that Baudrillard makes it an equation: information = entropy. Baudrillard argues that information destroys its base. The reason for this is twofold. First, information is always *about* something; it isn't that thing or experience itself. Information by definition, then, is always something other than the thing itself. Second, anytime we convey information, we must use a medium, and it is impossible to put something through a medium without changing it in some way. Even talking to your friend about an event you experienced changes it. Some of the meaning will be lost because language can't convey your actual emotions, and some meaning will be added because of the way your friend individually understands the words you are using.

Mass media is the extreme case of both these processes: Social information is removed innumerable times from the actual events, and capitalist mass media colors things more than any other form. One of the reasons behind this coloration is that mass media expends itself on the staging of information. Every medium has its own form of expression—for newspapers it's print and for television it's images. Every piece of information that is gleaned by the public from any media source has thus been selected and formed by the demands of the media. This is part of what is meant by the phrase "the medium is the message."

Further, mass communication comes prepackaged in a meaning form. What I mean by that is that information is staged and the subject is told what constitutes her or his particular relationship to the information. The reason for this is that media in postmodernity exist to make a profit, not to convey information. "Information" is presented more for entertainment purposes than for any intellectual ones. The concern in media is to appeal to and capture a specific market segment. That's why Fox News, CNN, and National Public Radio are so drastically different from one another—the information is secondary; its purpose is to draw an audience that will respond to appeals by capitalists to buy their goods. The presentation of information through the media is a system of self-referencing simulation or fantasy. Thus, "information devours its own content. It devours communication and the social" (Baudrillard, 1981/1994, p. 80).

There is yet another important factor in this decisive break—advertising: "Today what we are experiencing is the absorption of all virtual modes of expression into that of advertising" (Baudrillard, 1981/1994, p. 87). The act of advertising alone reduces objects from their use-value to their sign-value. For Marx, the movement from use-value to exchange-value entails an abstraction of the former; in other words, the exchange-value of a commodity is based on a representation of its possible uses. Baudrillard argues that advertising and mass media push this abstraction further. In advertising, the use-value of a commodity is overshadowed by a sign-value. Advertising does not seek to convey information about a product's use-value; rather, advertising places a product in a *field of unrelated signs* in order to enhance its cultural appearance. As a result of advertising, we tend to relate to the fragmented sign context rather than the use-value. Thus, in postmodern society, people purchase commodities more for the image than for the function they perform.

Let's think about the example of clothing. The actual use-value of clothing is to cover and keep warm. Yet right now I'm looking at an ad for clothing in *Rolling Stone,* and it doesn't mention anything about protection from the elements or avoiding public nudity. The ad is rather interesting in that it doesn't even present itself as an advertisement at all. It looks more like a picture of a rock band on tour. We could say, then, that even advertising is advertising itself as something. On the sidebar of the "band" picture, it doesn't talk about the band. It says things like "jacket, $599, by Avirex; T-shirt, $69, by Energie," and so on. Most of us won't pay $599 for a jacket to keep us warm; but we might pay that for the status image that we think the jacket projects. Baudrillard's point is that we aren't connected to the basic human reality of keeping warm and covered; we aren't even connected to the social reality of being in a rock band (which itself is a projected image of an idealized life). We are simply attracted to the images.

Simulacrum and Hyperreality

Baudrillard further argues that many of the things we do today in advanced capitalist societies are based largely on *images* from past lives. For example, most people in traditional and early industrial societies *worked with* their bodies. Today, increasing numbers of people in postindustrial societies don't work with their bodies, they *work out* their bodies. The body has thus become a cultural object rather than a means to an end. In postindustrial society, the body no longer serves the purpose of production; rather, it has become the subject of image creation: We work out in order to alter our body to meet some cultural representation.

Clothing too has changed from function to image. In previous eras, there was an explicit link between the real function of the body and the clothing worn—the clothing was serviceable with reference to the work performed or it was indicative of social status. For example, a farmer would wear sturdy clothing because of the labor performed, and his clothing indicated his work (if you saw him away from the field, you would still know what he did by his clothing). However, in the postmodern society, clothing itself has become the creator of image rather than something merely serviceable or directly linked to social status and function.

Let's review what we know so far: In times past, the body was used to produce and reproduce; clothing was serviceable and was a direct sign of work and social function. In postmodern society, however, bodies have been freed from the primary burden of labor and have instead become a conveyance of cultural image. The body's "condition" is itself an important symbol, and so is the decorative clothing placed upon the body. Now here is where it gets interesting: In the past, a fit body represented hard work and clothing signed the body's work, but what do our bodies and clothing symbolize today? Today, we take the body through a workout rather than actually working with the body. We work out so we can meet a cultural image—but what does this cultural image represent?

In times past, if you saw someone with a lean, hard body, it meant the person lived a mean, hard life. There was a real connection between the sign and what it referenced. But what do our spa-conditioned bodies reference? Baudrillard's point is that there is no real objective or social reference for what we are doing with our bodies today. The only reference to a real life is to that of the past—we used to have fit bodies because we worked. Thus, in terms of real social life, today's gym-produced bodies represent the past image of working bodies. Further, what does this imply about the clothes we wear? The clothes themselves are an image of an image that doesn't exist in any kind of reality. Baudrillard calls this **simulacrum,** an image of an image of a "reality" that never existed and never appears.

Thus, what we buy today aren't even commodities in the strictest sense. They are what Fredric Jameson calls free-floating signifiers. **Free-floating signifiers** are signs and symbols that have been cut loose from their social and linguistic contexts, and thus their meaning is at best problematic and generally nonexistent. In such a culture, tradition and family can be equated with paper plates (as in a recent television commercial) and infinite justice with military retribution (as in the U.S. president's initial characterization of the Iraq conflict).

Rather than representing, as signs did in the first phase, and rather than creating meaningful and social relations, as symbolic exchange did, commodified signs do nothing and mean nothing. They have no referent, and their sign-context, the only thing that could possibly impart meaning, is constantly shifting because of mass media and advertising. As they are, these signs cannot provoke an emotional response from us. Emotions, then, develop "a new kind of flatness or depthlessness, a new kind of superficiality in the most literal sense . . . [which is] perhaps the supreme formal feature of all the postmodernisms" (Jameson, 1984, p. 60).

These free-floating signs and images don't represent reality; they create hyperreality. Part of the hyperreality is composed of the commodities that we've been talking about. But a more significant part is provided by the extravagance of media entertainment, like Las Vegas and Disneyland. **Hyperreality** is a way of understanding and talking about the mass of disconnected culture. It comes to substitute for reality. In hyperreality, people are drawn to cultural images and signs for artificial stimulation. In other words, rather than being involved in social reality, people involve themselves with fake stimulations. Examples of such simulacra include artificial Christmas trees, breast implants, airbrushed Playboy Bunnies, food and drink flavors that don't exist naturally, and so on.

A clear example of this kind of hyperreality is reality television. Though predated by *Candid Camera*, the first reality show in the contemporary sense was *An American Family*, shown on Public Broadcasting Service stations in the United States. It was a 12-installment show that documented an American family going through the turmoil of divorce. The show was heavily criticized in the press. Today, reality shows are prevalent. Though the numbers are difficult to document, between the year 2000 and 2005 there were some 170 new reality shows presented to the public in the United States and Great Britain, with the vast majority being shown in the United States. The year 2004 saw reality programming come of age, as there were nine reality shows nominated for 23 Emmy awards.

The interesting thing about reality programming is that there is no reality. However, it presents itself as a *representation* of reality. For example, the show *Survivor* placed 16 "castaways" on a tropical island for 39 days and asked, "Deprived of basic comforts, exposed to the harsh natural elements, your fate at the mercy of strangers . . . who would you become?" (*Survivor* Show Concept, n.d.). But the "castaways" were never marooned nor were they in any danger (as real castaways would be). And the game rules and challenges read more like *Dungeons and Dragons* than a real survivor manual, with game "challenges" and changes in character attributes for winning (like being granted "Immunity"). So, what do the images of reality programming represent? Perhaps they are representations of what a fantasy game would look like if human beings could really get in one. The interesting thing about reality programming, and fantasy games for that matter, is the level of involvement people generate around them. This kind of involvement in simulated images of non-reality is hyperreality.

The significance of the idea of hyperreality is that it lets us see that people in postmodernity seek stimulation and nothing more. Hyperreality itself is void of any significance, meaning, or emotion. But, within that hyperreality, people create unreal worlds of spectacle and seduction. Hyperreality is a postmodern condition, a virtual world that provides experiences more involving and spectacular than everyday life and reality.

Sign Implosion

Baudrillard characterizes the simulacrum and hyperreality of postmodernity as an implosion. Where in modernity there was an explosion of signs, commodities, and distinctions, postmodernity is an implosion of all that. In modernity, there were new sciences like sociology and psychology; in postmodernity, the divisions between disciplines have collapsed along with an increasing preference for and growth of multidisciplinary studies. In modernity, there were new distinctions of nationality, identity, race, and gender; in postmodernity, these distinctions have imploded and collapsed upon themselves. Postmodernity is fractal and fragmented, with everything seeming political, sexual, or valuable—and if everything is, then nothing is. Baudrillard claims that this implosion of signs, identities, institutions, and all firm boundaries of meaning have led to the end of the social.

What Baudrillard is saying is that the proliferation, appropriation, and circulation of signs by the media and advertising influence the condition of signs, signification, and meaning in general. In the first stage of the sign, signs had very clear and specific meanings. But as societies and economic systems changed, different kinds of media and ideas were added. In the postmodern age, the media used to communicate information become utterly disassociated from any kind of idea of representation. Everywhere a person turns today in a postmodern society there are media. Cell phones, computers, the Internet, television, billboards, "billboard clothing" (clothing hocking brand names), and the like surround us. And every medium is commodified and inundated with advertising. In postmodernity, we are hard pressed to find any space or any object that isn't communicating or advertising something beyond itself.

All of this has a general, overall effect. Signs are no longer moored to any social or physical reality; all of them are fair game for the media's manipulation of desire. Any cultural idea, image, sign, or symbol is apt to be pulled out of its social context and used to advertise and to place the individual in the position of consumer. As these signs are lifted out of the social, they lose all possibility of stable reference. They may be used for anything, for any purpose. And the more media that are present, and the faster information is made available (like DSL versus dialup computer connections), the faster signs will circulate and the greater will be the appropriation of indigenous signs for capitalist gain, until there remains no sign that has not been set loose and colonized by capitalism run amok. All that remains is a yawning abyss of meaninglessness—a placeless surface that is incapable of holding personal identity, self, or society.

Let's take a single example—gender. Gender has been a category of distinction for a good part of modernity. Harriet Martineau, the first person to ever use the word "sociology" in print, saw some 80 years before women won the right to vote that the project of modernity necessarily entailed women's rights. So close is the relationship between the treatment of women and the project of modernity for

Martineau (1838/2003), that to her it becomes one of the earmarks of civilization: "Each civilized society claims for itself the superiority in its treatment of women" (p. 183). When gender first came up as an issue of equality, specifically in terms of a woman's right to vote, there was little confusion about what gender and gender equality meant. It was common knowledge who women were and what that very distinct group wanted.

But as modernity went on, things changed. In the United States after the 1960s, the single category of gender broke down. Various claims to distinction began to emerge around gender: The experience of gender is different by race, by class, by sexuality, and so on. The category imploded, with all the implicit understandings that went along with it. Today, when confronted by a person who appears to be a woman, the observing individual may be unsure. Is the "woman" really a woman or a man trapped in a woman's body? Or, is the "woman" a transsexual, who has physically been altered or symbolically changed (as with someone like RuPaul)? In postmodernity, the given cues of any category or object or experience cannot be taken at face value as indicative of a firm reality. All of the signs are caught up in a whirlwind of hyperreality.

In postmodernity, very few if any things can be accepted at face value. Meaning and reality aren't necessarily what they appear, because signs have been tossed about by a media without constraint, driven by the need to squeeze every drop of profit out of a populace through the proliferation of new markets with ever-shifting directions, cues, signs, and meanings in order to present something "new."

I've pictured my take of Baudrillard's argument in Figure 15.1. Let me emphasize that my intention with this diagram is simply to give you a heuristic device; it's a way to order your thinking about Baudrillard's theory. One of the best things about postmodern theory is that it is provocative, and part of the reason that it's provocative is that it isn't highly specified. There's a way, then, that this kind of picture stifles the postmodern; there's also a way in which something like this is quite modernist: It seeks to reduce complexity by making generalizations. So, I present this figure with some trepidation, but with the intent of giving you a place to "hang

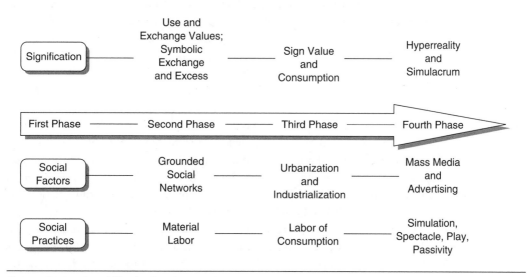

Figure 15.1   Baudrillard's Sign Stages

your hat." (In other words, if you're feeling at all lost and uncomfortable, then you need a modernist moment.)

As you can see, I've divided Baudrillard's thought into three main groups: signification, social factors, and social practices. I see Baudrillard as weaving these themes throughout his opus. And I've organized these ideas around the notion of sign phases. Generally speaking, the first two phases of the sign were fairly well grounded in the social. People were clearly involved in face-to-face social networks, the principal form of labor was material, commodities basically contained use- and exchange-values, and humanity was deeply entrenched in symbolic exchange.

The third phase of the sign comes about as the result of capitalism and the nation-state. The strong social factors in this era were urbanization and industrialization; people began to spend increasing amounts of time doing consumption labor; and the value of commodities shifted from use- to sign-value. As communication and transportation technologies increased, mass media and advertising became the important factors in social change. In a world of pure sign, signs no longer signify; it's all hyperreality and simulacrum, the fourth stage of the sign.

Concepts and Theory: The Postmodern Person

Fragmented Identities

Baudrillard envisions the "death of the subject." The subject he has in mind is that of modernity—the individual with strong and clear identities, able to carry on the work of democracy and capitalism. That subject, that person, is dead in postmodern culture. With the increase in mass media and advertising, there has been a corresponding decrease in the strength of all categories and meanings, including identities. What is left is a mediated person, rather than the subject of modernity. As Kenneth Gergen (1991) puts it, "one detects amid the hurly-burly of contemporary life a new constellation of feelings or sensibilities, a new pattern of self-consciousness. This syndrome may be termed *multiphrenia,* generally referring to the splitting of the individual into a multiplicity of self-investments" (pp. 73–74, emphasis original). This is an important idea, and one that appears in the work of a number of postmodernists. So let's take a moment to consider it fully.

One of the fundamental ways in which identity and difference are constructed is through exclusion. In psychology, this can by and large be taken for granted: I am by definition excluded from you because I am in my own body. For sociologists, exclusion is a cultural and social practice; it's something we *do,* not something we *are.* This fundamental point may sound elementary to the extreme, but it's important for us to understand it. In order for me to be me, I can't be you; in order for me to be male, I can't be female; in order for me to be white, I can't be black; in order to be a Christian, I can't be a Satanist; ad infinitum.

Cultural identity is defined in opposition to, or as it relates to, something else. Identity and self are based on exclusionary practices. The stronger the practices of exclusion, the stronger will be the identity; and the stronger my identities, the stronger will be my sense of self. Further, the greater the exclusionary practices, the more real will be my experience of identities and self. Hence, the early social movements for equality and democracy had clear practices of exclusion. It gave the people the strength of identity to make the sacrifices necessary to fight.

The twist that postmodernism gives to all this is found in the ideas of cultural fragmentation and de-centered selves and identities. In many ways, the ideas of gender and race are modernist: They collapse individualities into an all-encompassing identity. However, a person isn't simply female, for example—she also has many identities that crosscut that particular cultural interest and may shift her perceptions of self and other in one direction or another. For us to claim any of these identities—to claim to be female, black, male, or white—is really for us to put ourselves under the umbrella of a grand narrative, stories that deny individualities in favor of some broader social category. Grand narratives by their nature include very strong exclusionary tactics.

More to the point, the construction of centered identities is becoming increasingly difficult in postmodernity. Postmodernists argue that culture in postindustrial societies is fragmented. Since culture and identity are closely related, if the culture is fragmented, then so are identities. The idea of postmodernism, then, makes the issues of gender and race very complex. Clear racial and gender identities may thus be increasingly difficult to maintain. The culture has become more multifaceted, and so have identities. This means that we have greater freedom of choice, which we think we enjoy, but freedom of choice also implies that the distinctions between gender and racial identities aren't as clear or as real as they once were. Thus, social movements around race and gender become increasingly difficult to produce in postmodernity.

How then are postmodern identities constructed? Zygmunt Bauman (1992) argues that as a result of de-institutionalization, people live in complex, chaotic systems. Complex systems differ from the mechanistic systems in that they are unpredictable and not controlled by statistically significant factors. In other words, the relationships among the parts are not predictable. For example, race, class, and gender in a complex system no longer produce strong or constant effects in the individual's life or self-concept.

Thus being a woman, for instance, might be a disability in one social setting and not have any meaning at all in another, and race, class, and gender might come together in a specific setting in unique and random ways. Within these complex systems, groups are formed through unguided self-formation. In other words, we join or leave groups simply because we want to. And the groups exist not because they reflect a central value system, as a modernist would argue; rather, they exist due to the whim and fancy of their members and the tide of market-driven public sentiment.

The absence of any central value system and firm, objective evaluative guides tends to create a demand for substitutes. These substitutes are symbolically, rather than actually or socially, created. The need for these symbolic group tokens results in what Bauman (1992, pp. 198–199) calls "tribal politics" and defines as self-constructing practices that are collectivized but not socially produced. These neo-tribes function solely as imagined communities and, unlike their premodern namesake, exist only in symbolic form through the commitment of individual "members" to the *idea* of an identity.

But this neo-tribal world functions without an actual group's powers of inclusion and exclusion. It is created through the repetitive and generally individual or imaginative performance of symbolic rituals and exists only so long as the rituals are performed. Neo-tribes are thus formed through concepts rather than actual social groups. They exist as imagined communities through a multitude of agent acts of self-identification and exist solely because people use them as vehicles of self-definition: "Neo-tribes are, in other words, the vehicles (and imaginary sediments) of individual self-definition" (Bauman, 1992, p. 137).

Play, Spectacle, and Passivity

All that we just covered is caught up in the social practices of postmodernity. Opposition is impossible because postmodern culture has no boundaries to push against; it is tantamount to pushing against smoke. Further, the acting subject is equally as amorphous. Thus, according to Baudrillard, there's little place in postmodernity to grab ahold of and make a difference. Most things turn out to be innuendo, smoke, spam, mistakes . . . a smooth surface of meaninglessness and seduction. Baudrillard (1993a) leaves us with a few responses: play, spectacle, and passivity.

What can you do with objects that have no meaning? Well, you can play with them and not take them seriously. "So, all that are left are pieces. All that remains to be done is to play with the pieces. Playing with the pieces—that is postmodern" (Baudrillard 1993a, p. 95). How do we play in postmodernity? What would postmodern play look like? Here's an example: People in postmodern societies intentionally engage in fleeting contacts. Consider the case of "flash mobs." According to Wikipedia.com, "A flash mob is a group of people who assemble suddenly in a public place, do something unusual or notable, and then disperse. They are usually organized with the help of the Internet or other digital communications networks" (Flash Mob, n.d.). Sydmob (n.d.), an Internet group facilitating flash mobs in Sydney, Australia, asks, "Have you ever been walking down a busy city street and noticed the blank look on people's faces? How about on public transport? That look of total indifference is unmistakable; it's the face of [a] person feeling more like a worker bee than a human being. Have you ever felt like doing something out of the ordinary to see their reaction?"

In this play, spectacle becomes important. Georg Simmel, a classic theorist ahead of his time, gives us the same insight. Simmel (1950) argues that "life is composed more and more of these impersonal contents and offerings that tend to displace the genuine personal colorations and incomparabilities. This results in the individual's summoning the utmost in uniqueness and particularization, in order to preserve his most personal cores. He has to exaggerate this personal element in order to remain audible even to himself" (p. 422). Echoing Simmel, Zygmunt Bauman (1992) notes that "to catch the attention, displays must be ever more bizarre, condensed and (yes!) disturbing; perhaps ever more brutal, gory and threatening" (p. xx).

The remaining postmodern practice is a kind of resistance through passivity—refusing to play. There's an old American slogan from the Vietnam era that says,

"Suppose they gave a war and nobody came?" This is similar to what Baudrillard has in mind with resistance through passivity. Rather than attempting to engage postmodern culture, or responding in frustration, or trying to change things, Baudrillard advocates refusal or passive resistance. And perhaps like the war that no one shows up for, postmodernity will simply cease.

Summary

- Baudrillard uses Marx's notion of use- and exchange-value to argue that commodities are principally understood in postmodernity in terms of their sign-value.
- Baudrillard proposes four stages of the sign. In the first two stages, the sign adequately represented reality, and social communities were held together through the reciprocity of symbolic exchanges. People were also able to practice "excess"—the boundless potential of humanity—through festivals, rituals, sacrifices, and so on.
- Modernity began in the third phase of the sign with the advent of capitalism and industrialization. Baudrillard characterizes modernity as the consumer society. Within such a society, labor shifts to techniques of consumption with an eye toward sign identification. Modernity also brought rationalization and constraint, the antithesis of symbolic exchange and excess.
- In postmodernity, the increasing presence and speed of mass media, along with ever-increasing levels of commodification and advertising, push all vestiges of meaning out of signs. Mass media tends to empty cultural signs because the natural entropy of information is multiplied and because signification is suppressed in favor of media concerns of production. Advertising pushes this process of emptying further: Advertising sells by image rather than use, which implies that commodities are placed in unrelated sign-contexts in order to fit a media-produced image. These detached and redefined images are pure simulacra. Postmodern society is inundated by media technology and thus an immense amount of this kind of signification and culture, most of which references and produces a hyperreality.
- Baudrillard's postmodern condition is found in simulation, spectacle, play, and passivity. Baudrillard claims that the central subject of modernity—the person as the nexus of national and economic rights and responsibilities—is dead. In the place of the subject stands a media terminal of fragmented images. What remains is play and spectacle. As signs move ever faster through the postmodern media, their ability to hold meaning continues to disintegrate. Thus, in order to make an impression, cultural displays must be more and more spectacular. In such a climate, the hyperreality of media becomes more enticing, with greater emotional satisfaction than real life—but these media images must continue to spin out to ever more radical displays. Playing with empty signs or intentionally disengaging are the only possible responses.

Building Your Theory Toolbox

Learning More—Primary Sources

- I suggest that you start off with some of Baudrillard's more theoretical works:
  - *For a critique of the political economy of the sign,* Telos Press, 1981.
  - *The mirror of production,* Telos Press, 1975.
  - *Simulacra and simulation,* University of Michigan Press, 1994.
- There is also a good reader out:
  - *Baudrillard: A critical reader,* edited by Douglas Kellner, Blackwell, 1994.

Learning More—Secondary Sources

- *Jean Baudrillard: From Marxism to postmodernism and beyond,* by Douglas Kellner, Stanford University Press, 1990.

Check It Out

- *What about Disneyland?* To find out about Baudrillard's take on Disneyland, either read *America* by Baudrillard, Verso, 1989; or type "Disneyland and Baudrillard" into your Internet search engine.

Seeing the World

- After reading and understanding this chapter, you should be able to answer the following questions (remember to answer them *theoretically*):
  - How does Baudrillard argue that Marx actually affirms and legitimates capitalism? How does Baudrillard invert Marx's argument?
  - Explain how ideology and fetishism are based on a passion for the code.
  - What are the four stages of the sign? Be certain to explain their characteristics and the social factors that helped bring them about.
  - What is the consumer society? What labor is specific to the consumer society?
  - How do mass media and advertising empty the sign of all meaning and reference?
  - How have social identities imploded in postmodernity? What are the ramifications for political change?
  - Explain why play, spectacle, and passivity make sense in Baudrillard's postmodernity.

Engaging the World

- Using your favorite search engine, type in nostalgia and postmodern. What place does nostalgia have in postmodernity? How does Baudrillard talk about it? Give at least three examples of how this part of postmodernity is affecting your life.

Weaving the Threads

- What are the implications if both Baudrillard and Luhmann are right? Think about these implications in terms of meaning, society, and the individual.

- Compare and contrast the idea of the death of the subject and post-modern identities (neo-tribes) with Giddens's theory of the reflexive project of the self. Which do you think more accurately describes the issues surrounding identities and selves in late- and postmodernity? Why? (Remember to justify your answer *theoretically.*)
- In what ways, if any, does the postmodern notion of simulacrum or free-floating signifiers push Derrida's ideas of text? Explain your answer either way.
- Compare and contrast the political implications of Giddens's late-modernity and Baudrillard's postmodernity. Which do you think more accurately reflects the current conditions? Why?

INTRODUCTION
TO SECTION IV

Identity Politics

With identity politics we come at last to C. W. Mills's final question of the sociological imagination: What varieties of men and women now prevail in this society and in this period? It appears that Mills had at least two things in mind when he penned this question. He first wanted to know what kind of people prevail, in the sense of succeed and triumph. Mills no doubt had in mind the "power elite" of the age. Yet Mills goes on and asks further questions, indicating how this one question might be approached. In these further questions he asks, "What kinds of 'human nature' are revealed in the conduct and character we observe in this society in this period? And what is the meaning of 'human nature' of each and every feature of the society we are examining?" (Mills, 1959, p. 7).

In phrasing his questions this way, I think that Mills is asking us to look beyond the power elite and those that prevail politically and economically. He moves us to consider "each and every feature" of society and to consider the meaning of human nature. In thinking about human nature, I want to enlist the help of Erving Goffman (1967): "Universal human nature is not a very human thing. By acquiring it, the person becomes a kind of construct, built up not from inner psychic propensities but from moral rules that are impressed upon him from without" (p. 45). According to Goffman, human nature is what humans make it to be at any given point in their history. It's a social construct—and that is precisely why Mills is interested in it. The way we see subjects, subjectivity, selves, and identities—all of which speak of human nature—tells a great deal about our society. Society can't change without our understanding of these things changing too.

We began this book by considering the self of symbolic interaction. There we saw the self as a social object, a position from which to view our own behaviors, to give them meaning, but, more importantly, to control them. Society is possible

CHAPTER 16

Race Matters

Cornel West (1953–)

Photo: © Richard Howard/Time Life Pictures/Getty Images.

The problem of the twentieth century is the problem of the color-line.

(Du Bois, 1903/1996b, p. 107)

The title of this chapter, Race Matters, gives us two different ways of thinking about race in the United States. The first is perhaps the more apparent when we think of the sociology of race—race *matters:* It is an important factor in determining the life chances of certain groups of people in the United States. Consider the following:

- The Tuskegee study: In 1932, the U.S. Public Health Service began a study of untreated syphilis in black men. A total of 399 men signed up; the men were never told that they had syphilis, just that they had "bad blood"; the men were never given treatment (even after the advent of penicillin in 1947); as men died, went blind, or went insane, treatment was still withheld; the study discontinued after becoming public in 1972—but by that time 128 men had died of syphilis or related complications, at least 40 wives had been infected with syphilis, and 19 children were infected at birth.
- The proportion of males in prison: For whites the proportion is 462 per 100,000 in the population; for blacks it is 3,500 per 100,000 in the population.

- Nearly twice as many blacks as whites are convicted of drug offenses, even though it is estimated that there are 5 times more white users than black.
- Of youths charged with drug use, blacks are 48 times more likely than whites to receive a prison sentence.
- Of all racial groups, blacks have the lowest life expectancy; highest infant mortality; highest rate of most cancers, diabetes, heart disease, high blood pressure, new cases of HIV and AIDS; and the highest death rate from treatable diseases, gunshots, and drug/alcohol-related illness or incidents.
- Average annual income for black households: $30,436; for whites: $44,232
- Black unemployment is twice as high as that for whites.
- 48% of blacks in the United States own their home; 74% of whites do.
- The National Cancer Institute spends a mere 1% of its budget on studies of ethnic groups, even though black women are 50% more likely than white women to get breast cancer before 35 and 50% more likely to die before 50.
- EPA study: An estimated 9 out of 10 of the major sources of industrial pollution are in predominately black neighborhoods. The town of Carville, Louisiana, is 70% black and produces 353 pounds of toxic material per person per year—the Louisiana state average is 105 pounds.

This list of facts makes it clear that race has been an important issue in shaping the lives of African Americans. However, William Julius Wilson (1980), one of the most important race theorists of our time, argues that race as a causal factor in determining the overall condition of blacks in the United States has been declining in significance since the civil rights era. Wilson delineates three distinct phases of black–white relations in the United States: the period of plantation economy and racial-caste oppression, pre– and early post–Civil War America; the era of industrial expansion and class-based racial oppression, the latter part of the 19th century through the New Deal of the 1930s; and the time of progressive transition from racial inequalities to class inequalities, the current period beginning after WWII and taking shape during the 1960s and 1970s.

In racial-caste oppression, race was a strong determinate of the social and economic positions of blacks in the United States, particularly in the Southern plantation economy. In such systems of production, the aristocracy dominates both economic and political life. In the United States, the white landed elite were able to secure laws and policies extremely favorable to their economic interests, and they were able to propagate a ruling ideology concerning the differences between the races. As in classic Marxism, the system of production and the state formed a mutually reinforcing cycle. As a result, "the system of slavery severely restricted black vertical and horizontal mobility" (Wilson, 1980, p. 24) and race relations with elite whites took the form of paternalism.

After the Civil War and up through the 1930s, race continued as an important element in oppressing blacks, though things changed drastically. After the Civil War, the industrialization of the U.S. economy grew quickly and the Southern economy in particular expanded rapidly. In addition, the 13th and 14th Amendments to the Constitution abolished slavery and granted civil rights to the black population, respectively. As a result, from the latter part of the 19th century through the 1930s, there were massive changes in the system of production and race relations. This period marks a shift from race relations based on caste to class-based oppression.

Specifically, what changed was the position of white workers and the relationship between the state and capitalist elites. The plantation era had allowed there to be a close connection between the bourgeoisie and the state. This fit the classic Marxian model. As Marx (1848/1978c) said, "the Bourgeoisie has at last . . . conquered for itself, in the modern representative State, exclusive political sway. The executive of the modern State is but a committee for managing the common affairs of the whole bourgeoisie" (p. 475). After the Civil War, the state's relationship with capitalism began to change, not only in the South but in the rest of the country as well, as a result of antitrust legislation and worker protection laws.

In addition, the position of white workers changed markedly. In the South, economic expansion greatly increased the political power of the white working class. Blacks were free but had very little economic or political power. White workers, then, attempted to monopolize the newly available skilled and unskilled positions. The outcome was an elaborate system of Jim Crow segregation that was reinforced with a strong ideology of biological racism.

Further, the initial support from the white elite that blacks enjoyed in the South disappeared once the politicians began to worry about the black vote. To alleviate their uncertainty, white politicians effectively eliminated the black vote through such prerequisites as poll taxes, literacy tests, and property qualifications. "The almost total subordination of blacks in the South was clearly related to the disintegration of the paternalistic bond between Negroes and the southern economic elite, because this disintegration cleared the path for what ultimately resulted in a united white segregation movement" (Wilson, 1980, p. 83).

A united front never materialized in the North. Due to high levels of migration of blacks from the South and high immigration rates of European whites, blacks entered the job market as strikebreakers. White workers would strike for better wages or working conditions, and management would bring in black workers to keep production going. In some cases, management would preempt a strike by hiring on black workers permanently. This move obviously created high tension between black and white workers, which culminated in a number of race riots in 1917 and 1919.

Notice there is a shift in the source of oppression. During the plantation economy, oppression came out of collusion between the state and white bourgeoisie. But during the second era of race relations in the United States, a split labor market developed where the bourgeoisie wanted to give blacks economic freedoms in order to use them to keep white worker wages down. Thus, while race was still an issue, it originated with white workers rather than collusion between capitalists and the state.

Wilson argues that the role of the state continued to change from the classic Marxian model. WWII brought a ban on discrimination in defense and government agencies. This move also provided for on-the-job training for blacks. Black workforce participation continued to expand under the equal employment legislation of the 1950s and 1960s and the development and growth of affirmative action programs. These changes obviously didn't come as a result of the government's desire for equality, but in response to civil rights movements, which also boasted black political involvement. But regardless of the source, the state took successive steps to address black inequality.

As a result of affirmative action, more and more businesses were seeking black employees. For example, during the 10-year period between 1960 and 1970, the average number of corporate recruitment visits to traditionally black colleges jumped from 4 to 297; in some Southern colleges, the number rose from zero to 600 corporate visits. During this time, there was also a jump in the percentage of blacks working in government jobs, rising from 13% to almost 22%; the overall percentage of black males in white-collar positions rose from 16% to 24%.

However, Wilson (1980) argues that these shifts have "had little significance for the occupational advancement of the black underclass" (p. 103). Most, if not all, "affirmative action programs have benefited those blacks who are able to qualify for the expanding white-collar salaried positions in the corporate sector, positions that have higher educational and training requirements" (p. 100). This, coupled with shifts in the economy and the movement of industry out of urban settings, has meant that the black underclass is locked into low-paying and dead-end jobs. Importantly for Wilson's work, these jobs do not tend to be subject to interracial conflict.

Thus, Wilson argues, the effect of race on the condition of the black population in the United States has diminished. Job placement for blacks today has more to do with class than race. In addition, this segmented labor market, with a gulf between the interests of middle-class and poor blacks, has meant that it is "increasingly difficult to speak of a single or uniform black experience" (Wilson, 1980, p. 144).

Both Wilson's argument and the factual examples found at the beginning of this chapter are concerned with the importance of race as a determining factor in the lives of African Americans. According to Wilson, race still matters, though its importance is declining in the face of class inequality. With Cornel West, we come to a different way of understanding "race matters." Here the issue isn't so much the importance of race, but, rather, the *matters* of race: the cultural and political ramifications of living as African American in the United States. While West certainly wouldn't deny the structural changes and their effects on blacks that Wilson points out, West sees the upward mobility of some blacks and the split between black classes in terms of market forces and culture rather than social structure and class.

Most segments of American society have now been identified as target markets in capitalist consumerism. But saturation of the African American market has led to unique effects on the *experience of black identity*. It is these with which West is most concerned. West asks, how have the changes in class and marketability significantly altered black subjectivity and identity? Being concerned with culture and political identity leads West to conclude that race continues to matter profoundly. One further note on West: He clearly sees race and democracy as related. True democracy cannot exist in the face of racism. As West (1999) says, "To be American is to raise perennially the frightening democratic question: What does the public interest have to do with the most vulnerable and disadvantaged in our society?" (p. xix).

The *Web Byte* for this chapter gives us yet another view of race. Patricia Hill Collins argues that it is inadequate to view race in isolation. Because people occupy more than one status position, they stand at a place of "intersectionality" where race, class, gender, sexuality, religion, age, and so on come together. Collectively, these systems of inequality form a "matrix of domination" in society, which captures the overall organization of power relations. Thus, Collins gives us a very nuanced look at how power, oppression, and identity come together. Check it out: *Patricia Hill Collins and Intersecting Oppressions.*

The Essential West

Biography

Cornel West was born in Tulsa, Oklahoma, on June 2, 1953. He began attending Harvard University at 17 and graduated 3 years later, magna cum laude. His degree was in Near Eastern languages and literature. West obtained his Ph.D. at Princeton, where he studied with Richard Rorty, a well-known pragmatist. West has taught at Union Theological Seminary, Yale Divinity School, the University of Paris, and Princeton. He is currently a university professor at Harvard.

Passionate Curiosity

Cornel West's life is committed to not only the race question in America, but to the democratic ideals of open and critical dialogue, the freedom of ideas and information, and compassion for and acceptance of diverse others.

Keys to Knowing

market saturation, black nihilism, market moralities, ontological wounds, black leadership crisis, politics of conversion, racial reasoning, moral reasoning, Constantinian Christianity, democracy, postdemocratic age, free-market fundamentalism, aggressive militarism, escalating authoritarianism, niggerization, gangsterization, democratic armor

---

West's Perspective: Prophetic Democracy

Critical Culture

For West (1999), Marxian thought is an "indispensable tradition for freedom fighters" (p. 212). West is particularly interested in Georg Lukács's expansion of the Marxian relationship between commodities and false consciousness. Lukács (1923/1971) developed Marx's ideas and argued that the process of commodification affects every sphere of human existence and is the "central, structural problem of capitalist society in all its aspects" (p. 83). Under these circumstances, value is determined not by any intrinsic feature of human activity or relationships, but rather by the impersonal forces of markets, over which individuals have no control. The objects and relations that will truly gratify human needs are hidden, and the commodified object is internalized and accepted as reality. Thus, commodification results in a consciousness based on reified, false objects. This Lukácsian "reified mind" does not attempt to transcend its false foundation, but through rationalization and calculation "progressively sinks more deeply, more fatefully and more definitively into the consciousness of man" (p. 93)—in other words, people caught in consumerism tend to justify unending buying by any means possible.

One of the classic Marxian problems has to do with overcoming false consciousness with class consciousness. For both Marx and Lukács, the only group of people who can overcome false consciousness is the working class. The only way to truly see the whole is by standing outside of it—thus, only the proletariat can conceive of the social system in its entirety: "The proletariat represents the true reality, namely the tendencies of history awakening into consciousness" (Lukács, 1923/ 1971, p. 199). The working class by its very position of alienation is capable of seeing the true whole, the knowledge of class relations from the standpoint of the entire society and its system of production and social relations. Bourgeois thought is simply an ahistorical acceptance of its own ideology and status quo.

Obviously, there is a tension here. On the one hand, commodification results in false consciousness, and on the other hand, the workers—those who suffer the effects of commodification the most—are the only ones who can see the problem for what it is. There is also another problem: the "overwhelming resources of knowledge, culture and routine which the bourgeoisie undoubtedly possesses" (Lukács, 1923/1971, p. 197). Thus, though the working and middle classes are capable of grasping the whole, they are susceptible both to psychological false consciousness and the cultural resources of the elite. However, these cultural resources are not controlled by the elite exclusively. West draws from another mid–20th-century Marxist, Antonio Gramsci (1971), to argue that the ability to rule does not depend on material relations alone, as in classic Marxian thought. Social change will involve a war of cultural positions— a battle for people's minds, not overt conflict.

Culture, then, takes on a critical value for West. Generally, this cultural position is important in race theory in two ways. First, it is the place of revolutionary work. As we've seen, according to this tradition of Marxian thinking, there is first a battle for the mind in any social change. There must be a critical intelligence. West sees this type of consciousness as part of the project of modernity itself, specifically the American democratic experiment. "My conception of what it means to be modern is shot through with a sense of the dialogical—the free encounter of mind, soul and body that relates to others in order to be unsettled, unnerved and unhoused" (West, 1999, pp. xvii–xviii). As we will see, this kind of culture doesn't just create itself, especially in capitalist countries. Critical awareness comes out of specific kinds of culture-producing practices, most notably community, religion, and public discourse.

The second way culture is important for race theory is because of its long heritage. Many of the early publications by people of color and women were in the form of literature. Harriet Beecher Stowe's *Uncle Tom's Cabin* and Charlotte Perkins Gilman's *The Yellow Wallpaper* are good examples. While the work of W. E. B. Du Bois is obviously scholastic, it is permeated with poems, songs, and literary references. For example, Du Bois' powerful *The Souls of Black Folk* begins with the lyrics and music notes of a song. Du Bois' other major work, *Darkwater,* begins with a credo, a confession of faith.

The choice on the part of these authors to produce literary work may have been due to the structural constraints placed upon political minorities, but it is also true that the plight of the disenfranchised often may be best communicated through means that can impact emotion and not simply cognition. Oppression isn't simply a fact; it is a profound experience. West (1999) points this out when he says that "our great truth tellers [are] mainly artists" (p. xix).

Further, the troubles themselves cry out for creative release. Note the way West (1999) describes music, the "highest expression" of human history: "Music at its best achieves this summit because it is the grand archaeology into and transfiguration of our guttural cry, the great human effort to grasp in time . . . our deepest passions and yearnings as prisoners of time. Profound music leads us—beyond language—to the dark roots of our scream and the celestial heights of our silence" (p. xvii). Now notice how Marx (1844/1978a) describes religion: "*Religious* suffering is at the same time an *expression* of real suffering and a *protest* against real suffering. Religion is the sigh of the oppressed creature, the sentiment of a heartless world, and the soul of soulless conditions. It is the *opium* of the people" (p. 54, emphasis original). The last part of that quote is what most people are familiar with, but taken in context it achieves a much fuller meaning. These intense forms of culture—music and religion—both articulate something about the human condition that is difficult to express elsewhere. This link between profound experience and deep cultural expression may be one reason why music and religion have occupied such prominent places in African American culture.

Philosophy: Pragmatism and Existentialism

Pragmatism

We discussed pragmatism in Chapter 1. But let's review it here. Briefly, pragmatism is the American philosophy that developed out of the cultural devastation of the Civil War as a way of understanding truth. Truth in pragmatism is specific to a community. It arises out of the collective's physical work and communication. Thus, pragmatism is based on common sense and the belief that "truth and knowledge shifts to the social and communal circumstances under which persons can communicate and cooperate in the process of acquiring knowledge" (West, 1999, p. 151).

Thus, in pragmatism, human action and decisions aren't determined or forced by society, ideology, or preexisting truths. Rather, decisions and ethics emerge out of a consensus that develops through interaction. This is the pragmatism that West (1999) latches onto. It is a culture of creative democracy . . . where politically adjudicated forms of knowledge are produced in which human participation is encouraged and for which human personalities are enhanced. Social experimentation is the basic form, yet it is operative only when those who must suffer the consequences have effective control over the institutions that yield the consequences. (p. 151)

Existentialism

I want to start our consideration of existentialism by asking you a question: Have you ever wondered why you exist, or what the meaning of life is? If so (and most of us have), you've experienced the essence of existentialism. Existentialism starts with the problem of being or existence and argues that the very question or problem *creates* existence. As far as we know, human beings are the only animal that questions its existence; all other animals simply exist. But human beings ask, and in asking we create human existence as a unique experience. Thus, the answer to the question can only be found in the being asking it.

Here's a simple analogy. Let's say I come to you with a cup of coffee and declare that it tastes like mud. And I then ask, "Why does this coffee taste like mud?" You taste the coffee and tell me that it tastes fine. "Okay," I say, "but why does it taste like mud?" Startled, you respond by telling me once again that the coffee doesn't taste like mud. And I insist, "Why does the coffee taste like mud?" Eventually, you would probably get frustrated and tell me that the muddy coffee is my problem, not yours. Why is it my problem? Because I'm the only one that thinks the coffee tastes like mud. Both the problem and its solution can only exist inside of me. The same is true for the question of existence.

Two ideas are prominent in existentialism: struggles or suffering, and authenticity or being (existence). These ideas find space between such tensions as the individual versus the social, subjective versus objective, liminal experiences (such as death) and nihilism, and situated thought versus reason. Whatever the tensions may be, the resolution is found within the person her- or himself. As we've seen, pragmatism argues that there is no absolute truth, only truths that are arrived at socially, truths that provide practical answers to real problems. Truth for pragmatists, then, is socially practical. In contrast to pragmatism, existentialism argues that there are no social truths and truth exists emotionally, not objectively. Truths such as history and science—social, practical truths—belong to society or to the crowd. As such, they are objective and indifferent. Truth in existentialism is determined by individual authenticity—by a passionate, and generally absurd, leap of faith.

This issue of subjective authenticity is exemplified by the biblical story of Abraham (See Søren Kierkegaard's *Fear and Trembling*). The story goes that one day God came to Abraham and commanded him to kill his son. For Abraham, there appeared to be little to justify such a command. In fact, Abraham had every reason to doubt that the command even came from God, since even Satan himself can appear as an angel of light. Even so, Abraham prepared to kill his son and offer his life on an altar of burning wood. With little or no objective evidence, what drove Abraham to make the sacrifice? Biblical believers would say faith, and existentialism

Black Cultural Armor

These two kinds of black leaders have promoted political cynicism among black people, and have dampened "the fire of enraged local activists who have made a difference" (West, 1993/2001, p. 68). Part of black nihilism, or nothingness, is this sense of ineffectuality, of being lost in a storm too big to change. What is needed, according to West, are black leaders founded on moral reasoning rather than racial reasoning.

Moral reasoning is the stock and trade of race-transcending prophetic leaders. Prophetic leadership does not rest on any kind of racial supremacy, black or white. It uses a coalition strategy, which seeks out the antiracist traditions found in all peoples. It refuses to divide black people over other categories of distinction and rejects patriarchy and homophobia. Such an approach promotes moral rather than racial reasoning.

This prophetic framework of moral reasoning is also based on a mature black identity of self-love and self-respect that refuses to put "any group of people on the pedestal or in the gutter" (West, 1993/2001, p. 43). Moral reasoning also uses subversive memory, "one of the most precious heritages [black people] have" (West, 1999, p. 221). It recalls the modes of struggling and resisting that affirmed community, faith, hope, and love, rather than the contemporary market morality of individualism, conspicuous consumption, and hedonistic indulgence.

Both the *coalition strategy* and mature black identity are built at the local level. West (1999) sees local communities as working "from below and sometimes beneath modernity" (p. 221), as if local communities can function below the radar of markets and commodification. It is within vibrant communities and through public discourse that local leaders are accountable and earn respect and love. Such leaders merit national attention from the black community and the general public, according to West.

In this framework, the liberal focus on economic issues is rejected as simplistic. Likewise, the conservative critique of black immorality is dismissed as ignoring public responsibility for the ethical state of the union. In their places, West proposes a democratic, pragmatically driven dialogue. As I mentioned earlier, West doesn't propose absolutes. His is a prophetic call to radical democracy and faith, to finally take seriously the declaration that all people are created equal.

Together, moral reasoning, coalition strategy, and mature black identity create the black cultural armor. West's use of "armor" is a biblical reference. Christians are told in Ephesians 6:13 (New International Version) to "put on the full armor of God, so that when the day of evil comes, you may be able to stand your ground, and after you have done everything, to stand." There the threat was the powers of darkness in heavenly places; here the threat is black nihilism in the heart of democracy. These two battles are at least parallel if not identical for West. The fight for true democracy is a spiritual battle for the souls of humankind that have been dulled by market saturation, especially the souls of black America. West (1993/2001) exhorts black America to put on its cultural armor—a return to community life and moral reasoning along with coalition strategy and mature black identity—so as to "beat back the demons of hopelessness, meaninglessness, and lovelessness" and create anew "cultural structures of meaning and feeling" (p. 23).

Concepts and Theory: Postdemocratic Age

West's (2004) newest work is an indictment of American democracy in the wake of 9/11. He argues that the terrorist attack on September 11, 2001, provided the spark to an already existing fire bed of antidemocratic dogmas and an emasculated political process. West notes that the political scene in the United States has recently been dominated by an illicit marriage of corporate and political elites (a plutocracy) and the Christian Right. Among the plutocratic elite, "salesmanship to the demos has taken the place of genuine democratic leadership" (p. 3). Given the choice between two political alternatives that are both dependent upon corporate favor, people are increasingly choosing to opt out of the democratic process, both in terms of voting and critical dialogue. West characterizes this as a **postdemocratic** age—"the waning of democratic energies and practices in our present age of the American empire" (p. 2).

West (2004) sees the emptiness of the American political culture, created by market saturation and a dearth of leadership, as giving place to the Christian Right. People are reaching out for a sense of meaning and purpose. The Christian Right provides that, but its righteousness is misguided and its perspective "narrow, exclusionary, and punitive" (p. 66). Specifically, the rhetoric of Christian fundamentalism is used to legitimate three antidemocratic dogmas: free-market fundamentalism, aggressive militarism, and authoritarianism. As Berger and Luckmann (Chapter 8) argue, religion provides the strongest legitimation of any belief or idea—any fundamentalism is the most powerful force behind any religious legitimation. In addition, West (2004) argues that the Christian Right is perverting the soul of American democracy, "because the dominant forms of Christian fundamentalism are a threat to the tolerance and openness necessary for sustaining any democracy" (p. 146).

Yet, as we noted in the discussion about West's perspective, West is a Christian. But he sees vast differences between what he calls Constantinian Christianity and prophetic Christianity. *Constantinian Christianity* is named after the Roman emperor Constantine, who converted to Christianity in 312 C.E. The various accounts differ on some of the specifics, but all agree that Constantine received a vision of Christ just before the Battle of the Milvian Bridge. As a result, Constantine commanded that a purple silk banner hanging from a crosspiece on a pike (representing Christ) be placed as his new battle standard. Eventually, because of Constantine, Christianity became the official religion of the Roman Empire. The state then used the church as an instrument of imperial policy, and the church used the state as a means of imposing its religious rule. Constantinian Christianity, then, is a "terrible co-joining of church and state" (West, 2004, p. 148) that robs the church "of the prophetic fervor of Jesus and the apocalyptic fire of that other Jew-turned-Christian named Paul" (p. 147).

West (2004) argues that as a result of the marriage between church and state, Christianity has been invested with an "insidious schizophrenia." On the one hand, there are the Constantinian elements occupied with power, privilege, and possession, a Christianity that has "been on the wrong side of so many of our social troubles, such as the dogmatic justification of slavery and the parochial defense of women's inequality" (p. 149). West argues that the Christian Right, including the Christian Coalition and the Moral Majority, is the shining example of Constantinian Christianity in America.

On the other hand are the elements of prophetic Christianity, most clearly seen in Social Gospel churches. Prophetic Christianity holds up wisdom, justice, and freedom for all humanity as its virtues. It isn't concerned with power; it is concerned with promoting equality and respecting and supporting every cultural group's unique heritage and way of life. In making his case for the differences between Constantinian and prophetic Christianity, West (2004) argues that the strongest movements for equality have been led by prophetic Christians, including "the abolitionist, women's suffrage, and trade-union movements in the nineteenth century and the civil rights movement in the twentieth century" (p. 152).

> I do not want to be numbered among those who sold their souls for a mess of pottage—who surrendered their democratic Christian identity for a comfortable place at the table of the American empire while, like Lazarus, the least of these cried out and I was too intoxicated with worldly power and might to hear, beckon, and heed their cries. To be a Christian is to live dangerously, honestly, freely. . . . This is the kind of vision and courage required to enable the renewal of prophetic, democratic Christian identity in the age of the American empire. (West, 2004, p. 172)

Three Antidemocratic Dogmas

Before we begin this section, we should take a moment to define what West means by democracy. *Democracy* is not simply the freedom to vote—the freedom to vote democratically is based on the presence of at least three elements. Together, these elements give democracy a forward vision—the hope of future progress gained through rejecting the shackles of the past and the continual process of enlightenment. The first element of democracy is open and critical dialogue. The democracy of the United States is built upon such dialogue, as is evident in the Declaration of Independence and the First Amendment to the Constitution. The second element is necessitated by the first: the freedom of ideas and information necessary for democratic dialogue and questioning. Democracy cannot exist in an environment where knowledge and thought are hidden in darkness.

Third, the necessity of dialogue and the freedom of ideas imply compassion for and acceptance of diverse others. A democratic government exists in order to preserve the freedoms and rights of diverse others. Any kind of government can protect its borders and provide infrastructure, but West argues that a democratic government is especially well-suited to guard the freedoms of its citizens in the face of oppression. This protection is the defining feature of a democratic government and its sole reason for existence. Note that acceptance is not the same as tolerance. Tolerated voices aren't allowed an equal footing in dialogue. But American democracy goes further than acceptance. In the roots of American democracy there is desire for alternative voices. Consider these words from "The New Colossus" by Emma Lazarus:

Not like the brazen giant of Greek fame
With conquering limbs astride from land to land;
Here at our sea-washed, sunset gates shall stand
A mighty woman with a torch, whose flame
Is the imprisoned lightning, and her name
Mother of Exiles. From her beacon-hand
Glows world-wide welcome; her mild eyes command
The air-bridged harbor that twin cities frame,
"Keep, ancient lands, your storied pomp!" cries she
With silent lips. "Give me your tired, your poor,
Your huddled masses yearning to breathe free,
The wretched refuse of your teeming shore,
Send these, the homeless, tempest-tossed to me,
I lift my lamp beside the golden door!"

As we've seen, West takes seriously the idea that culture can exist and act like a structure. This position implies first that culture can develop autonomously and second that culture can have its own set of effects in concert with or independent of other social structures. In this case, the social structural issues that concern West are the rising plutocracy and the Christian Right. West is also still concerned with the saturation of market forces. In addition, West sees the terrorist attacks of 9/11 as a key event in pushing the United States toward a post-democratic society.

There are three cultural dogmas with which West is concerned: free-market fundamentalism, aggressive militarism, and escalating authoritarianism. We'll talk about each of these in a moment, but first notice West's use of religious terms. First, these cultural issues are *dogmas*. While dogma can have a more general meaning, in religious circles it is a technical term with a very specific meaning. Dogmas are officially established religious doctrines. They serve not only to distinguish one belief system from another, but they are also the guiding lights for religious practice. West is telling us that these cultural elements function like religious dogma: They dictate and legitimate certain beliefs and practices. And these beliefs, practices, and legitimations are held to be fundamental to a certain way of life—in this case, the American way of life.

The second religious term that West uses is *fundamentalism,* and it is used in reference to the first cultural belief: free markets. Interestingly enough, though we may now talk about Islamic fundamentalists, the term was first used in reference to Protestant Christianity. Christian fundamentalism began in the United States during the 19th century in response to several millennialist movements (belief in the second coming and 1,000-year reign of Christ). It spread during the latter part of the 19th century because of Protestant concerns over Catholic immigration, labor unrest, and biblical criticism. There are several beliefs that are common to fundamentalism, including the literal interpretation of the Bible, the physical second coming of Christ, physical resurrection, and so on. But *what* people believe is not as important as *how* people believe—fundamentalism is characterized by absolute certainty and militant conservatism.

In terms of free markets, West (2004) is arguing that American culture has developed a militant belief in them—a *free-market fundamentalism*. Free markets are perceived to be the mechanism for bringing about international cooperation, modernization, happiness, true competition, the good life, and so forth. This glorification of the market leads to a corporate-dominated society where the interests of capitalism are equated with democracy and corporate leaders are seen as the highest expression of democratic good. The current idea of free markets seems to be shielded by a faith that borders on worship; little is done in the face of the estimated $300 billion cost of white collar crime (Legal Information Institute, n.d.). Market fundamentalism also "redefines the terms of what we should be striving for in life, glamorizing materialistic gain, narcissistic pleasure, and the pursuit of narrow individualistic preoccupations"; in the end, it "trivializes the concern for public interest" (p. 4).

West's (2004) point is that this dogmatic belief in free-market fundamentalism blinds people to any other concern except market moralities. Because of this single-sightedness, individuals caught up in this belief are willing to make any sacrifice necessary to "succeed at any cost" (p. 27). The irony of this blind chase for profit is that it is based on faith in free markets, which implies that even "corporate elites are not fully in control of market forces" (p. 27) and market forces are a power unto themselves. The effects of unfettered market forces are beginning to show in terms of environmental pollution and the emptying out of democratic energies.

Part of the drain on democratic energies comes from nihilism. Just as West (2004) saw black nihilism as a result of the saturation of market forces and market moralities, he also argues that Americans as a whole are suffering from "psychic depression, personal worthlessness, and social despair" (p. 26). Nihilism comes about because (1) people are caught in the ambivalence of believing in free markets while recognizing their inevitable costs (such as white collar crime); (2) unceasing market expansion and market moralities create a situation where meaning and knowledge are continually overturning and are stripped of any continuity or stability; (3) market moralities are void of any sense of right or wrong (which is the definition of moral); and (4) market moralities have corrupted the principles of representative democracy all the way to the top:

> Our politicians have sacrificed their principles on the altar of special interests; our corporate leaders have sacrificed their integrity on the altar of profits; and our media watchdogs have sacrificed the voice of dissent on the altar of audience competition. (West, 2004, p. 28)

The next anticapitalist dogma that West explains is *aggressive militarism*. Although the roots of both this and the next dogma (authoritarianism) go deeper, the immediate stimulus is the terrorist attacks of 9/11. West (2004) sees 9/11 as a watershed moment for American democracy: "The ugly events of 9/11 should have been an opportunity for national self-scrutiny" (p. 12). Why did the terrorists attack the United States? What ideas are so meaningful that men would not only kill but lay down their own lives to communicate? What are the social, political, and historical backgrounds? Did U.S. imperialist behavior play a role? But rather than becoming positively responsive, West points out, the citizens of this nation either killed any hope of communication in a "simplistic and aggressive 'with us or against us' stance" (p. 12) or were silenced by the single-note dogmatism of American fundamentalism.

Thus, rather than democratic questioning, the United States turned to the dogma of aggressive militarism. The new national policy of war became defined in terms of the "preemptive strike," where in the presence of even faulty intelligence it is in the nation's best interest to attack first. In practice, this policy is evidenced by unilateral intervention, colonial invasion, and armed occupation of foreign soil and people. West clarifies the costs of such actions: At the international level, the use of unmasked violence tends to create further instability; at the national level, the dogma of militarism results in expanded police power, further development of the prison-industrial complex, and increased legitimation of male power and violence; at the ethical level, such elite-driven war actions are always paid for most dearly in the disproportional deaths of youth coming from lower classes and populations of color.

The third member of this trinity of anticapitalistic dogma is *escalating authoritarianism*. West argues that the American belief in authoritarianism is rooted in the understandable paranoia about terrorism, the longstanding fear of individuals having too many liberties, and the deep-seated distrust of any social or cultural differences. The events of 9/11 rekindled and deepened these fears. The results have gone in the exact opposite direction from democracy: dramatic increases in governmental surveillance, especially within schools and universities, and decreases in legal protections of individual citizens—these two are coupled with and amplified by decreases in the oversight of government activities.

West sees an analogy here. In America, he sees black people as having been "niggerized." To be *niggerized* is to be dehumanized, and blacks in this country have been designated and treated as "nigger" for over 350 years. They have been made "to feel unsafe, unprotected, subject to random violence, and hated" (p. 20). In short, blacks have been terrorized by white America. Now, the entire United States has been niggerized. As a result of the terrorism of 9/11, Americans in general feel unsafe, unprotected, subject to random violence, and hated. The comparison West wants to make is between the black response to terrorism and the current American response to terrorism.

West (1999) characterizes the current response to 9/11 as the *gangsterization* of America. A gangster mentality is one that makes things a "question of getting over . . . instead of getting better, and that gangster mentality promotes a war against all" (p. 218). This gangsterization of America not only concerns aggressive militarism and escalating authoritarianism, it also involves market moralities and fetishes that come out of free-market fundamentalism. These dogmas have produced "an unbridled grasp at power, wealth, and status" that have snuffed the democratic light from the very nation that is its chief advocate—"we are experiencing the sad American imperial devouring of American democracy" (West, 2004, p. 8).

Putting on Democratic Armor

To combat the nihilist threat, to overcome the niggerization and gangsterization of America, West exhorts us to put on the democratic armor. There are three

music forms in America, yet they were created by an enslaved people. Blues specifically gives expression to tragicomic hope.

The blues originated in the back-and-forth call of slaves working the fields. The call at once gave voice to pain and hope to the soul. Grief was expressed in the call, yet the cadence gave rhythm and thus lightness to the work. Individual suffering was expressed, affirmed, and given meaning in a community of sufferers. "The root of blues is the human experience and psyche itself" (Erlewine, 1999, p. v), and its essence is "to stare painful truths in the face and persevere without cynicism or pessimism" (West, 2004, p. 21).

Tellingly, the blues was born out of terrorism: the terrorist suppression of blacks by white supremacist slave owners. West tells us that we are at a crossroads brought about by yet another kind of terrorism: the attacks of 9/11. According to West (2004), we have begun down the wrong road, toward a postdemocratic society. Even so, "our fundamental test may lie in our *continuing* response to 9/11" (p. 8, emphasis added). America's move toward the road of democracy begins with the blues:

> The blues forges a mature hope that fortifies us on the slippery tightrope of Socratic questioning and prophetic witness in imperial America. . . . This kind of tragicomic hope is dangerous—and potentially subversive—because it can never be extinguished. . . . It is a form of elemental freedom that cannot be eliminated or snuffed out by any elite power. (pp. 216, 217)

Summary

- West draws on four traditions in organizing his thought: cultural Marxism, pragmatism, existentialism, and Christianity. Marx's theory of commodification and market expansion are key elements in West's theory as well. But West also draws upon the critical Marxian view of culture—culture is used to oppress in often hidden ways, but it can also be used to bring change. From pragmatism, West takes the idea of practical values emerging from collective dialogue. He sees pragmatism as a key philosophy in the American democratic experiment. From existentialism, West draws the ideas of subjective authenticity and passionate absurdity. And from Christianity, he takes love for self and others and the prophetic outcry for mercy and justice.
- Since the 1960s, blacks in the United States have on the one hand enjoyed increased economic and political freedoms, but on the other have become the victims of market saturation. Market saturation has changed the primary orientations of blacks. Previously, blacks were strongly oriented to civic and religious institutions and the traditional ties of family and home. Market saturation has infested the black community with market moralities: fleeting hedonistic pleasure and monetary gain. The effects of markets are exaggerated for blacks because of the black heritage in America. The mix of past wounds, the continuing racial prejudice, and market moralities creates black nihilism (a sense of hopelessness and meaninglessness associated with living as a black person in the United States).

- West exposes a crisis in black leadership, arguing that most black leaders either fall under the managerial/elitist model or that of protest leaders. With protest leaders, racial reasoning is paramount, which promotes ethics based on skin color alone, rather than on moral or justice issues. West calls on prophetic leaders that will transcend race and return to moral reasoning. These leaders must begin in the community, at the grassroots level, where they can participate in pragmatic community dialogue, build up trust, and maintain accountability.
- West argues that since 9/11, America has entered a postdemocratic age. There are three dogmas that have worked to bring this about: free-market fundamentalism, aggressive militarism, and escalating authoritarianism. To return to democracy, West argues that we must put on the democratic armor: Socratic questioning, prophetic justice, and tragicomic commitment to hope.

Building Your Theory Toolbox

Learning More—Primary Sources

See the following works by Cornel West:
- *Race matters,* Vintage Books, 1993.
- *Democracy matters: Winning the fight against imperialism,* Penguin Books, 2004.
- *The Cornel West reader,* Basic *Civitas* Books, 1999.

Check It Out

- *Web Byte—Patricia Hill Collins and Intersecting Oppressions*
- Cornel West has put out a hip-hop CD. Find information at http://www.cornelwest.com/

Seeing the World

- After reading and understanding this chapter, you should be able to answer the following questions (remember to answer them *theoretically*):
  o West takes very specific things from Marxian theory, pragmatism, existentialism, and Christianity. Explain what those are and how they influence his work.
  o How can culture act as a structure?
  o What are the three structural forces influencing blacks in America today?
  o Why did market saturation affect the black community in unique ways?
  o What is black nihilism, where does it come from, and how is it affecting blacks in the United States today?
  o Why is there a black leadership crisis? What are politics of conversion?
  o What is racial reasoning? How does moral reasoning counter racial reasoning?

o What is West's critique of the Christian Right? How is the Christian Right an example of Constantinian Christianity?

o What are the three elements of democracy?

o How are free-market fundamentalism, aggressive militarism, and escalating militarism creating a postdemocratic age?

o How has the United States as a whole been niggerized and gangsterized?

o What is the democratic armor? How will each piece help overcome the postdemocratic age?

o How does West use the concepts of jazz and blues?

Engaging the World

- Become involved in campus efforts to end discrimination. Check and see if you have an office of multicultural affairs. Find out what other campus organizations are involved in ending discrimination.

Weaving the Threads

- Though the issues are different—gender versus race—compare and contrast Janet Chafetz's more scientific approach with West's political identity approach. Which do you think is more effective and why?

CHAPTER 17

Gendered Consciousness

Dorothy E. Smith (1926–)

Photo: Courtesy of Dorothy E. Smith.

My research concern is to build an ordinary good knowledge of the text-mediated organization of power from the standpoint of women in contemporary capitalism.

(Smith, 1992, p. 97)

I remember as a child being fascinated with the inner lives of other people. I would stare at an airplane as it flew overhead and I would continue to watch until it was out of sight, all the while wondering who was in there, where they were going, what they were doing, and how they were thinking and feeling. I would do the same with people in cars. The nice thing about my car-people was that they were visible, where my airplane-people were more mysterious because of their lack of visibility. Since I could see my car-people, I could thus put a category on them—they were male, female, black, white, old, or young. But the categories only made me more curious. These were *real* people, when compared to my airplane-people. I knew that because I could see them. But for all that I could see, I was still blind. And that blindness made me even more curious. Where is *she* going? What is *she* thinking? What's *her* home life like? Do the people in her life make her happy or sad?

All these questions and more would run through my head in an instant. I would feel connected to the other person but at the same time utterly foreign. I would want to follow her home and shadow her all through her day. The problem is that when I actually was around other people, following them wasn't enough. I then wanted to get inside their heads. I could see their lives; I could see who they interacted with and how they used the physical objects around them. But what was all that like to them? A woman may have seen the same person that I did, but how did she experience that person?

But I outgrew that fascination. Experientially, I'm not sure when it happened or where it went, but I realized one day that I don't look at people like that anymore. Most of the time people are just a blur to me, unless I'm focused on one individual. When I am focused on a person, it is usually for a rational, strategic reason (I need to buy groceries from the checker, or you need to talk about your grade with me), and I understand that individual within the roles and expectations that are important for the interaction. And, sadly, planes, trains, and automobiles are no longer fascinating for me—I don't wonder about the diverse lives of the people within them. They are simply commonplace objects that I am only dimly aware of as they move in and out of my field of reference.

While I may not know when or how my understanding changed, Dorothy Smith gives us a reasonable and provocative way to think about this. Smith isn't interested in the phases of childhood development, but she is centrally concerned with the differences in the points of view found in my story. The first perspective in my story is one that subjectivizes the individual. It grants and gives validity to the experiences of the individual within specific circumstances. The second perspective is one that objectivizes the individual. It only recognizes people as instances of social categories within general types of situations. If Smith were to give an explanation of my shift in perspectives, she probably wouldn't attribute it simply to childhood development or a busy life. In explaining the differences in perspectives, she would probably do so in terms of my being co-opted by the "relations of ruling."

The Essential Smith

Biography

Dorothy E. Smith was born in Northallerton, Yorkshire, Great Britain, in 1926. She earned her undergraduate degree in 1955 from the London School of Economics. In 1963, Smith received her Ph.D. from the University of California at Berkeley. She has taught at Berkeley, the University of Essex, and the University of British Columbia and is currently a professor emeritus at the University of Toronto, where she has been since 1977. In recognition of her contributions to sociology, the American Sociology Association (ASA) honored Smith with the Jessie Bernard Award in 1993 and the Career of Distinguished Scholarship Award in 1999. Her book *The Everyday World as Problematic* has received two awards from the Canadian Sociology and Anthropology Association: the Outstanding Contribution Award and the John Porter Award, both given in 1990.

Passionate Curiosity

Smith's passion is found at the intersection of text and life. Smith argues that the social and behavioral sciences have systematically developed an objective body of women that sets the problems and questions of research and provides the answers and theory. "Inquiry does not begin within the conceptual organization or relevances of the sociological discourse, but in actual experience as embedded in the particular historical forms of social relations that determine that experience" (Smith, 1987, p. 49).

Another way to put this issue is that most social research assumes a reciprocity of perspectives. One of the things that ethnomethodology (see Chapter 3) has taught us about the organization of social order at the micro level is that we all assume that our way of seeing things corresponds fairly closely to the way other people see things. More specifically, we assume that if another person were to walk in our shoes, they would experience the world just like we do. This is an assumption that allows us to carry on with our daily lives. It lets us act as if we share a common world, even though we may not and we can never know for sure. According to Smith, social science usually works in this way too, but she wants us to problematize that assumption in sociology. She wants us to ask, "What is it like to be *that* person in *that* body in *those* circumstances?"

Sociology and the Relations of Ruling

Smith (1990) talks about the practices, knowledge, and social relations that are associated with power as relations of ruling. Specifically, *relations of ruling* include "what the business world calls *management*, it includes the professions, it includes government and the activities of those who are selecting, training, and indoctrinating those who will be its governors" (p. 14, emphasis original). In technologically advanced societies that are bureaucratically organized, ruling and governing take place specifically through abstract concepts and symbols, or text. As Michel Foucault explained, knowledge is power; it is the currency that dominates our age. Authority and control are exercised in contemporary society through different forms of knowledge—specifically, knowledge that objectifies its subjects.

The social sciences in particular are quite good at this. They turn people into populations that can be reduced to numbers, measured, and thus controlled. Through abstract concepts and generalized theories, the social sciences empty the person of individual thoughts and feelings and reduce her or him to concepts and ideas that can be applied to all people grouped together within a specific social type. The social sciences thus create a textual reality, a reality that exists in "the literature" outside of the lived experience of people.

Much of this literature is related to data that are generated by the state, through such instruments as the U.S. Census or the FBI's Uniform Crime Reporting (UCR) Program. These data are accepted without question as the authoritative representation of reality because they are seen as *hard data*—data that correspond to the assumptions of science. These data are then used to "test" theories and hypotheses that are generated, more often than not, either from previous work or by academics seeking to establish their names in the literature. Even case histories that purport to represent the life of a specific individual are rendered as documents that substantiate established theoretical understandings.

Thus, most of the data, theory, and findings of social science are generated by a state driven by political concerns, by academics circumscribed by the discipline of their fields, by professors motivated to create a vita (résumé) of distinction, or by professionals seeking to establish their practice. All of this creates "textual surfaces of objective knowledge in public contexts" that are "to be read factually. . . . as evidences of a reality 'in back of' the text" (Smith, 1990, pp. 191, 107). Therefore, a sociology that is oriented toward abstract theory and data analysis results in a sociology that "is a systematically developed consciousness of society and social relations. . . . [that] claims objectivity not on the basis of its capacity to speak truthfully, but in terms of its specific capacity to exclude the presence and experience of particular subjectivities" (Smith, 1987, p. 2).

These concepts, theories, numbers, practices, and professions become *relations of ruling* as they are used by the individual to understand and control her own subjectivity, as she understands herself to be *a subject of* the discourses of sociology, psychology, economics, and so on. We do this when we see ourselves in the sociological articles or self-help books we read, in the written histories or newspapers of society, or in business journals or reports. With or without awareness of it, we mold ourselves to the picture of reality presented in the "textual surfaces of objective knowledge."

Smith points out that this process of molding becomes explicit for those people wanting to become sociologists, psychologists, or business leaders. Disciplines socialize students into accepted theories and methods. In the end, these are specific guidelines that determine exactly what constitutes sociological knowledge. For example, most of the professors you've had are either tenured or on a tenure track. Whether an instructor has tenure or not is generally the chief distinction between assistant and associate professors. And when a sociology professor comes up for tenure and promotion, one of the most important questions asked about her or his work is whether or not it qualifies as sociology. Not everything we do is necessarily sociology—it has to conform to specific methodologies, assumptions, concepts, and so on to qualify as sociology.

There is something reasonable about this work of exclusion. If I wrote an article with nothing but math concepts in it, it probably shouldn't be considered sociology. Otherwise there wouldn't be any differences among any of the academic disciplines. However, Smith's point is that there is more going on than simple definitions. Definitions of methods and theory are used by the powerful to exclude the powerless. What counts as sociology and the criteria used to make the distinctions are therefore reflections of the relations of ruling. Sociology and all the social sciences have historically been masculinist enterprises, which means that what constitutes sociology is defined from the perspective of ruling men. The questions that are deemed important and the methods and theories that are used have all been established by men: "How sociology is thought—its methods, conceptual schemes, and theories—has been based on and built up within the male social universe" (Smith, 1990, p. 13).

Let me give you an example to bring this home, one that has to do with race, but the illustration still holds. In the latter part of the 1990s, two colleagues and I were and nuanced and it can and will inspire intricate and subtle thought and research. But her point is rather straightforward—social research and theory need to be grounded in the actual lived experiences of people, particularly women.

The first thing I'd like for you to notice about Figure 17.2 is the central position of both actual lived experience and text. We've talked about the issue of lived experience, and I will come back to it in a minute, but let's start by noticing texts. As I've already noted, sociology and the social disciplines in general have experienced what has been called a "linguistic turn." A good deal of this sea change can be credited to the work of Jacques Derrida, poststructuralism, and postmodernism in general. We can understand the linguistic turn as a shift toward the importance of text, primarily written words, though other kinds of cultural artifacts such as film are read as texts through semiotic analysis. In this perspective, culture and cultural readings are fundamentally important. The radical thread in this linguistic turn is that readings of texts are themselves seen as texts, which means that since humans are defined through meaning, all we have are texts.

In keeping with the linguistic turn, Smith (1992) acknowledges the importance of text, but she adds to it the ontology of lived experience. For her, text forms "the bridge between the actual and discursive. It is a material object that brings into actual contexts of reading a fixed form of meaning" (p. 92). The uniqueness of what Smith is arguing in light of the linguistic turn is that there is something other than text. Text isn't everything: There are embodied people who live their lives in actual situations that have real consequences.

When we become aware of the texts that surround our lived reality, they form the bridge that Smith is talking about. There are two ways through which these texts can influence us. First, we may become directly aware of them, generally through higher education but also through the media. At this point, the discursive text directly enters the everyday life of people. This kind of text is generally authoritative; it claims to be the voice of true knowledge gained through scientific or organizational inquiry. However, as Smith points out, social scientific research is based outside of actual lived experience. Its position outside is in fact what makes this knowledge appear legitimate, at least in a culture dominated by scientific discourse. It is this appearance that prompts us to privilege the objective voice above our own. But there is more to these texts, as you can see from the left side of Figure 17.2.

The relations of ruling have a reciprocal relationship with social scientific inquiry, as noted by the double-headed arrow. We believe that legitimate research produces the only real knowledge; government finances, directs, and thus defines the kinds of research that are seen as legitimate. Social scientific inquiry then produces the kinds of data and knowledge that reinforce and legitimate the ruling. The single-headed arrow from relations of ruling to texts implies the top-down control of knowledge that Marx spoke of: The ruling ideas come from the ruling people, in this case men. The arrow between scientific inquiry and texts, however, is two-headed. This means that the questions and theories that social scientific research uses come from the literature rather than the real lives of people. It also implies that social science is in a dialogue with itself, between its texts and its inquiry.

The second way we can become aware of these texts is through social scientific inquiry itself. Have you ever answered the phone and found that someone wanted you to respond to a survey? Or have you ever been stopped in a mall and "asked a few questions" by someone with a clipboard? Have you ever filled out a census survey? Through all these ways and many others, we are exposed to objectifying texts by social scientific inquiry.

Notice that the arrow coming from social scientific inquiry has only one head, going toward the actual world of women, and notice that the arrow has two nouns: readers and objects. This one-way arrow implies that social scientific research produces readers and objects. The readers are the researchers. They are trained to read or impose their text onto the actual world. They see the lived experience of women through the texts and methods of scientific research. They come to real, actual, embodied life with a preexisting script, one that has the potential to blind them to the actualities of women. Further, when the questions and methods of science are used to understand women, women are made into objects, passive recipients of social sciences' categories and facts.

The right side of the model depicts Dorothy Smith's approach. There are two important things to notice. First, there are no relations of ruling controlling standpoint inquiry. Part of this is obvious. As I've mentioned, Smith says that this way of seeing things is applicable to all types of people, but it is particularly salient for women. The reason for its importance for women is that the relations of ruling are masculine in a society such as ours. Men control most of the power and wealth and thus control most of the knowledge that is produced. And while there is a difference between objective knowledge on the one hand and the lived experience of men on the other, women mitigate that discrepancy.

But this issue of ruling isn't quite that clear cut for Smith. Relations of ruling are obviously associated with men. However, there is a not-so-obvious part as well. The work of women, including feminists, can fall prey to the same problem that produces social scientific inquiry. This can happen when women reify the ideas, ideology, or findings of feminist research. Any time research begins outside of the lived experience of embodied people, it assumes an objective perspective and in the end creates abstract knowledge. This is how women's movements "have created their own contradictions" (Smith, 1992, p. 88). It's possible, then, for women's knowledge to take on the same guise as men's. In Smith's approach, there are no relations of ruling, whether coming from men or women. Standpoint inquiry must continually begin and end in the lived experiences.

The other thing I'd like to call your attention to is that all the arrows associated with standpoint inquiry are double-headed. Rather than producing readers and objects, standpoint inquiry creates space for translators and subjects. In standpoint, the lives of women aren't simply read; they aren't textually determined. A researcher using standpoint inquiry is situated in a never-ending dialog with the actual and textual. There is a constant moving back and forth among the voice of the subject, the voice of authoritative text, and the interpretations of the researcher. Smith (1992) sees this back-and-forth interplay as a dialectic: "The project locates itself in a dialectic between actual people located just as we are and social relations, in which we participate and to which we contribute, that have come to take on an existence and a power over [sic] against us" (pp. 94–95).

Notice that the dialectic is between actual experience and social relations. Smith is arguing that in advanced bureaucratic societies, our relationships with other people are by and large produced and understood through text. For example, you have a social relationship with the person teaching this class. What is that relationship? To state the obvious, the relationship is professor–student. Where is that relationship produced? You might be tempted to say that it is produced between you and your professor, but you would be wrong, at least from Smith's point of view. The relationship is *practiced* between you and your professor, but it is *produced* in the university documents that spell out exactly what qualifies as a professor and a student (remember, you had to apply for admittance) and how professors and students are supposed to act.

This textuality of relationships is a fact of almost every single relationship you have. Of course, the relations become individualized, but even your relationship with your parents (How many books on parenting do you think are available?) and with the person you're dating (How many articles and books have been written about dating? How many dating-related surveys have you seen in popular magazines?) are all controlled and defined through text. However, as we've already seen, Smith argues that even in the midst of all this text, there is a reality of actual, lived experience. Smith is explicitly interested in the dialectic that occurs between abstract, objectifying texts on the one hand, and the lived actualities of women on the other.

We thus come to the core of Smith's project. Recently (2005), Smith has termed this project "institutional ethnography." The "ethnography" portion of the term is meant to convey its dependence upon lived experience. Smith's project, then, is one that emphasizes inquiry rather than abstract theory. But, again, remember that Smith isn't necessarily arguing against abstractions and generalizations. Smith herself uses abstractions. Notice this quote from Smith (1987) concerning the fault line: "This inquiry into the implications of a sociology for women begins from the discovery of a point of rupture in my/our experience as woman/women within the social forms of consciousness" (p. 49). In it she uses both abstractions and particulars: my/our, woman/women. To say anything about women—which is a universal term—is to already assume and use a theoretical abstraction. Thus, Smith uses abstract concepts, so she isn't saying that in and of themselves they are problematic—the issue is what we do with them. Her concern is for when abstractions are reduced to "a purely discursive function" (Smith, 1992, p. 89). This happens when concepts are reified or when inquiry begins in text: "To begin with the categories is to begin in discourse" (p. 90).

There are, I think, two ways that Smith uses and approaches abstractions. First, in standpoint inquiry, concepts are never taken as if they represented a static reality. Lived experience is an ongoing, interactive process in which feelings, ideas, and behaviors emerge and constantly change. Thus, the concepts that come out of standpoint inquiry are held lightly and are allowed to transform through the never-ending quest to find out "how it works."

The second and perhaps more important way that Smith approaches theoretical concepts is as part of the discursive text that constitutes the mode through which relations of ruling are established and managed. As we've seen, "the objectification of knowledge is a general feature of contemporary relations of ruling" (Smith, 1990, p. 67). A significant principle of standpoint inquiry is to reveal how texts are put together with practices at the level of lived experience. "Making these processes visible also makes visible how we participate in and incorporate them into our own practices" (Smith, 1992, p. 90) and how we involve ourselves in creating forms of consciousness "that are properties of organization or discourse rather than of individual subjects" (Smith, 1987, p. 3).

It's at this point that Smith's use of the word "institutional" is relevant. It signals that this approach is vitally concerned with exploring the influences of institutionalized power relations on the lived experiences of their subjects. Institutional ethnography is like ethnomethodology and symbolic interactionism in that it focuses on how the practical actions of people in actual situations produce a meaningful social order. But neither of these approaches gives theoretical place to society's ruling institutions, as Smith's method does. In that, it is more like a contemporary Marxian account of power and text. Thus, **institutional ethnography** examines the dialectical interplay between the relations of ruling as expressed in and mediated through texts, and the actual experiences of people as they negotiate and implement those texts.

Smith uses the analogy of a map to help us see what she is getting at. Maps assist us to negotiate space. If I'm in a strange city, I can consult a map and have a fair idea of how to proceed. Maps, however, aren't the city and they aren't our experience. Smith (1992) wants sociology to function like a map—a map that gives an account of the person walking and finding her or his way (lived experience) through the objective structures of the city (text). This kind of sociology "would tie people's sites of experience and action into accounts of social organization and relations which have that ordinarily reliable kind of faithfulness to 'how it works'" (p. 94).

Specifically, Smith is interested in finding out just how the relations of ruling pervade the lives of women. These relations, as we've seen, come through texts and researchers. But in most cases, the relations of ruling are misrecognized by women. They are rendered invisible by the normalcy of their legitimacy. Part of what these maps can do, then, is make visible the relations of ruling and how they impact the lived experiences of women.

Smith is also interested in how actual women incorporate, respond to, see, and understand the texts that are written from a feminist or standpoint perspective. This is an important issue. Looking at Figure 17.2, we might get the impression that standpoint inquiry automatically and always produces translators and subjects. Another way to put this is that it appears as if standpoint inquiry is a static thing, as if, once done, the inquiry stands as the standpoint forever. This is certainly not what Smith is arguing. Notice again that double-headed arrow between standpoint and texts. Once standpoint inquiry is expressed in text, there is the danger that it will be taken as reality and become discursive. Smith's is thus an ongoing and ever-changing project that takes seriously the objectifying influence of text.

For me, then, the standpoint of women locates a place to begin inquiry before things have shifted upwards into the transcendent subject. Once you've gone up there, settled into text-mediated discourse, irremediably stuck on the reading side of the textual surface, you can't peek around it to find the other side where you're actually *doing* your reading. You can reflect back, but you're already committed to a standpoint other than that of actual people's experience. (Smith, 1992, p. 60, emphasis original)

Summary

- Smith argues that in contemporary society, power is exercised through text. Smith defines text using three factors: the actual words or symbols, the physical medium, and the materiality of the text. It is the last of the three with which Smith is most concerned, the actual practices of writing and reading. Most, if not all, of the texts produced by science, social science, and organizations achieve their facticity by eliminating any reference to specific subjectivities, individuals, or experiences.

- These texts are gendered in the sense that men by and large constitute the ruling group in society. Men work and live in these texts and thus accept them as taken-for-granted expressions of the way things are. Women's experience and consciousness, on the other hand, are bifurcated: They experience themselves within the text, as the ruling discourse of the age, but they also experience a significant part of their lives outside of the text. And it is in this part of women's lives where the contingencies of actual life are met, thus giving these experiences a firmer reality base than the abstract ruling texts of men. Further, men are enabled to take objective ruling texts as true because women provide the majority of the labor that undergirds the entire order.

- The bifurcated consciousness becomes particularly problematic for those women trained in such disciplines as business, sociology, psychiatry, psychology, and political science. In these professions, women are trained to write and read ruling texts, ignoring the lived experiences of women at the fault line.

- Smith proposes a theoretical method of investigation (standpoint inquiry or institutional ethnography) that gives priority to the lived experiences of women. In this scheme, texts are not discounted or done away with, but they are put into the context of the embodied, actual experiences of women. Smith thus opens up a site of research that exists in the dialectic interplay between text and women's experience.

Building Your Theory Toolbox

Learning More—Primary Sources

See the following works by Dorothy Smith:
- *The everyday world as problematic: A feminist sociology,* Northeastern University Press, 1987.
- *The conceptual practices of power: A feminist sociology of knowledge,* Northeastern University Press, 1990.
- *Institutional ethnography: A sociology for people,* AltaMira Press, 2005.

Learning More—Secondary Sources

- *Feminist theory: From margin to center,* by bell hooks, South End Press, 2nd edition, 2000.
- *Gender inequality: Feminist theories and politics,* by Judith Lorber, Roxbury Publishing, 3rd edition, 2005.

Check It Out

- *Web Byte—Patricia Hill Collins and Intersecting Oppressions*
- *Explore feminism:* http://www.amazoncastle.com/feminism/ecocult.shtml

Seeing the World

- After reading and understanding this chapter, you should be able to answer the following questions (remember to answer them *theoretically*):
  o Explain how standpoint is more method than theory. How does some feminist work actually defeat standpoint?
  o How are the relations of ruling expressed through social science?
  o What is the new materialism? How does it affect what people accept as true or factual?
  o Explain the differences between the general sociological approach and Smith's.
  o How does the fault line perpetuate gender inequality?
  o Describe Smith's institutional ethnography. How is it dialectical? What do you think the benefits of institutional ethnography would be?

Engaging the World

- As a student, how do you see yourself being socialized to the relations of ruling? If you are a woman, do you see bifurcated consciousness in your life? As a sociologist, how will you avoid being trapped and controlled by the relations of ruling?

Weaving the Threads

- Compare and contrast Foucault's and Smith's ideas about how power is mediated.
- What does Derrida's idea of poststructuralism say about Smith's text-mediated power?
- Compare and contrast Smith's idea of the fault line and Chafetz's theory of male micro-resource power.

Concepts and Theory: Bodies That Matter

Bringing the Body In

For quite some time, sociology's approach to gender has been to grant a kind of independent existence to the body and sex. This approach is reflective of how social thinkers in general have thought about the body, if they thought about it at all. For example, George Herbert Mead (1934) gives us a warning about putting too much attention on the body: "It may be necessary again to utter a warning against the easy assumption . . . that the body of the individual as a perceptual object provides a center to which experiences may be attached" (p. 357). This warning from Mead is particularly important for us to see because of all the classic or original thinkers in sociology, he would be the most likely to consider the body of consequence. Yet here he warns against giving too much significance to the body. Talcott Parsons was one of the first to explicitly include the physical body in theory, but for him the body was simply a biological organism—he made a distinction between organic and psychological systems in the hierarchy of control.

When social thinkers did begin to think about the body as perhaps something more than biology, it was as a social object, a thing that can take on symbolic meanings. Thus, Chris Shilling (1993), in one of the first sociological books dedicated to the body, says, "Growing numbers of people are increasingly concerned with the health, shape, and appearance of their own bodies as expressions of individual identity" (p. 1). Notice how Shilling talks about the body: It's a thing or social object with which we are increasingly concerned. This same approach to the body is seen in Seymour Fisher's (1973) book *Body Consciousness:* "My major intent in this book is to consider the strategies that people use in learning how to make sense of their own bodies" (p. ix). Again, the body is something separate from the actual person. It's a thing that we can decorate, be concerned about, or make sense of.

Pierre Bourdieu was one of the first theorists to clearly see that the body isn't simply an organism or a social object or even a vehicle for expression. Bourdieu argued that the body is "classed" through habitus. In Bourdieu's scheme, our bodies become encultured—they are socialized into class-based tastes and dispositions. Class is thus structured and replicated in our bodies.

Butler takes even Bourdieu's approach to the body an additional step. She tells us that the distinction between sex and gender is inaccurate. I'm sure you remember from any course on gender, even your introduction to sociology course, that your instructor said that sex and gender are different. The story goes that sex refers to the biological plumbing (the body) and gender to the socially constructed roles that are built around bodily sex differences. The reason for this division is to show that gender is a cultural entity, something that has been socially created to control and make distinctions between the sexes (the already existing body). It's a kind of ploy where sociology says, "Sure, there is biology in the form of sex, but it doesn't really matter. What matters is gender." In response, Butler says no, sex matters fundamentally.

Inscribing the Body

Butler argues that it isn't simply the practices associated with sex that are regulated, as Foucault argued; sex itself is controlled. Butler means sex as we usually mean it when we say sex and gender—sex as the biological condition of the body. However, Butler (1993) argues that we mistakenly think that sex only comes from biology: "The category of 'sex' is, from the start, normative. . . . In this sense, then, 'sex' not only functions as a norm, but is part of a regulatory practice that produces the bodies it governs, that is, whose regulatory force is made clear as a kind of productive power, the power to produce—demarcate, circulate, differentiate—the bodies it controls" (p. 1).

Using Derrida's idea of writing, Butler is arguing that bodies are inscribed. Derrida argued that arche-writing is a form of violent etching, engraving, or carving. As an analogy, we can think of the way young lovers used to carve their initials on trees. Derrida said that everything humans do is like that—we carve a significant reality upon the face of an otherwise smooth and meaningless surface. Butler is arguing that sex is just such an engraving, as if the body is a formless mass of senseless cells that society sculpts into what we call sex. The etching is violent because it forcibly denies other possibilities.

Notice that from the very beginning, sex is normative. Let's be very clear about something here. When we are talking about sex, we are not talking about "having sex." It isn't the activity that interests us here. You *are* a sex—a principal part of your very existence as a human is sex. And that sex is normative. What we mean by this is that sexed bodies are the result of social control. It's important to see that Butler isn't just talking about sexual behaviors or identities. Behaviors and identities are things that are *about* the body, but are not the body itself. It's not that things having to do with the body aren't important; they are. The way you relate to your body is crucial. You can tattoo it or not; you can condition it or not; you can abuse it through drugs and alcohol or not. In fact, the body is part of the reflexive project of the self in late modernity, according to Anthony Giddens. That's possibly one of the reasons why we are so concerned with it in our daily lives.

Butler (1993) is telling us that bodies matter intrinsically, in and of themselves— what they are and how they exist; their materiality. They matter because they are not simply biological organisms. Inside and out, the body is a social production— sex "is a regulatory ideal whose materialization is compelled, and this materialization takes place (or fails to take place) through certain highly regulated practices" (p. 1). Sex is what qualifies a body as a body. Try to imagine a human body without sex, and you probably come up blank. Sex is one of the essentials that qualify our bodies as specifically human—according to Butler (1993), sex is "that which qualifies a body for life within the domain of cultural intelligibility" (p. 2). We can't make sense of our bodies apart from sex.

Butler argues that the way the body is sexed is through psychodynamic processes surrounding the hegemonic norm of heterosexuality. That's a mouthful, so let's break it down. Let's take the easy part first: the **hegemonic norm of heterosexuality.** Hegemony means having superior influence or authority. Thus, what Butler is saying is that the superior authority over sex is heterosexual: two body types, male and female, that are mutually and exclusively attracted to one another. The key word in that definition is "exclusively." Of course, it is biologically necessary for the males and females of our species to come together sexually. But it is not necessary for dimorphic (two bodies) sexual attraction to be exclusive, as Foucault demonstrated in his histories of sexuality. For example, in Greek society it was normal for men to be sexually active with boys, yet the Greeks replicated the species just fine.

The psychodynamic process is a bit more difficult to understand. Anytime we are talking about psychodynamic theory, we are alluding to or basing our ideas upon Sigmund Freud. Freud argued that people develop an ego through psychosexual stages and the resolution of the Oedipus complex. For Freud, there are three parts to a person's inner being: the id, ego, and superego. The ego mediates the demands of the id and superego. Though there are important differences, the ego generally corresponds to what sociologists call the self.

People aren't born with either an ego or superego; at birth there is only the **id**, and it is the seat of our passions and desires. The important thing about the id is that it is the source of all our psychic energy, also known as the libido. It is our basic motivation in the world and corresponds to instinctual energy—the closest thing to instincts that humans have. The id pushes us to gratify our basic needs; among the most fundamental of these is sexual gratification. But the id bumps up against problems in its search for gratification. All its needs can't always be gratified immediately. This is obviously true for animals as well; the chief difference is that the basis of human gratification is other people and society. Thus, as the id meets resistance from its human environment when it cannot satisfy all of its urges, its energy kind of splits in two different directions: the ego and superego.

According to Freud, one of the most important steps in the development of the child is the Oedipus complex and its resolution. This stage occurs between the ages of 3 and 5. At the beginning of this phase, the child is fundamentally attached to the mother, most notably through breast-feeding and nurturing. This primary attachment occurs through the id and its libido energy, which means it is primarily sensual and emotional rather than intellectual. In order to successfully develop psychically, both boys and girls need to resolve this attachment issue. One of the challenges at this stage, especially for boys, is to detach from the mother and attach to the same-sex parent (for girls, this means reattaching to the mother), yet at the same time deny that sensual-emotional attachment. Resolution is achieved when the child identifies with the parent of the same sex and simultaneously represses its sexual instincts.

Remember, when babies are born they are nothing but a ball of desire. All of these desires are felt similarly and without distinction about when, why, how, or with whom the desires are met. Part of our development, then, is the channeling of these desires into their "proper" paths. A fair portion of this libido attachment that the child has to his or her mother is sexual. Freud didn't necessarily mean sexual in the way you or I might understand it. It's much more basic and primitive—part of the child's general instinctual drives. We can think of it like hunger for food. We have a fundamental instinct to eat. But what, when, and how we eat is culturally programmed. Psychodynamic theory sees sex in the same way. Sex doesn't magically appear at puberty; it's been there all along. It is a primary form of physical, sensual attachment to others. And, as with food, the child's desires must be channeled to "appropriate" objects of sensual attachment.

Of course, sex and food are different things. Sex is far more important for the psyche. With food there isn't a psychic perception of loss, just a sense of bodily hunger if we don't eat. But with sex there is a sense of psychic loss. Desires for the mother must be suppressed by the child in order to attach successfully to others. Notice how this idea of suppression parallels Derrida's notion of difference (*différance*)—significance is created by suppression, in this case suppressing general sensual attachment in favor of a socially defined one. Within the hegemonic norm of heterosexuality, this means that the boy needs to shift his sexual attachment to other women and the girl needs to transfer it to men. But the suppressed desires still exist in the id or the unconscious. It's the job of the superego and ego to keep these passions in check by denying them.

Thus, Butler draws on Freud but she gives him a Foucauldian twist. Foucault argued that all knowledge is historically specific and is best understood in terms of discourse. What this means is that Freud's understanding of the Oedipus complex is specific to this time—it isn't an essential stage that humans universally have to work through. And, more importantly, it has a *discursive function*. Here discursive has the sense of analytical reasoning or logic. According to Butler, the cultural logic behind the embodied sex of this age is heterosexism. Freud's theory occurs within a "prevailing truth-regime of 'sex'" (Butler, 1993, p. 233). The proper or appropriate heterosexual objects of sexual desire occur only within and simply because of this truth-regime.

Performity

Our bodies are thus inscribed and sculpted by a culturally specific sex. Sex isn't mere biology. In fact, there isn't anything that is simply biology for humans; we inscribe everything. That's how things become intelligible (and thus possible) for us. In this truth-regime of sex, our possibilities are heterosexual. Our heterosexism is inscribed upon our bodies during early childhood. Through psychodynamic processes, certain options are closed off (Derrida's *différance*), the body becomes intelligible and possible, and the power of hegemonic heterosexuality is reflexively applied by the individual (Foucault).

However, heterosexuality isn't a once-and-for-all accomplishment. The desires of the id are always present. That's why we need the superego. The *superego* is formed as the voice of society or the parent, most notably the father in Freud's scheme. It protects the ego from the unconscious and overwhelming demands of the id. Just as Derrida argued, some possibilities are held off in favor of others. Thus, our homosexual desires are continually repressed so that heterosexuality can be continuously inscribed upon our bodies.

Here Butler uses another idea from Derrida: *iterability*. We are most familiar with this word in a different form. We may say something like, "To reiterate, let me state . . ." Reiterate obviously means to say again. But "iterate" doesn't mean to say the first time. If you look up both reiterate and iterate in the dictionary, you will find that they both mean to repeat. And that is Derrida's point. In order to be intelligible, all words, all text, must be in principle repeatable.

Let's imagine that you have a word that is only going to be said once in all of time. You say that word. However, if that word cannot be understood, then it isn't a word at all, just sound. But if the word is understood by others, then even if it is never said aloud again, it must in principle be possible for the word to be repeated. The word's very intelligibility demands that it be repeatable. In fact, when you say the word and others hear it and understand it, that word is already being repeated in their minds. Every iteration is thus a reiteration, every statement is a restatement, and every text is a text of a text, according to Derrida.

To get at this idea for the sexed body, Butler uses the term **performity:** a word, phrase, or action that brings something into existence. A clear example of performity is the phrase "I now pronounce you" in a wedding ceremony; the phrase itself brings something into existence at the moment of its pronouncement. To understand performity a little better, let's talk about Erving Goffman's notion of impression management. Goffman argued that we perform or display gender for others. We construct a front through our use of appearance, manner, and setting that communicates to others the kind of gendered self we are claiming. Once effectively claimed by the self, other people impute the gendered self to the person. They act in a particular way toward the social category that has been claimed, and they expect the individual to live up to the self that has been presented. This self, then, is a dramatic enactment; it's something that we produce that allows interactions to take place. The effect of this self upon the individual comes primarily from other people: Others form righteously imputed expectations that the individual must live up to. When performances and expectations are repeated over time, the individual becomes attached to her or his "face."

Goffman's notion of performance thus gives a great deal of latitude to the performer. Butler doesn't see it like that. Individuals must perform or, better, iterate/reiterate sex. As I mentioned earlier, to be a body means to be sexed. Butler is talking about something that is much more fundamental than Goffman's gender display. She's talking about existing as a body. Other people don't simply understand us as a gender; they see us as a material being, as a person with a body. Our most important method of being embodied isn't what or how we eat, it isn't how we dress or decorate the body, it isn't our health level. According to Butler, we are principally embodied in and through sex. In particular, we are embodied through the performity of sex.

We need to take this one step further. Goffman claimed that others form righteous expectations, that these expectations are attributed to us, and that we then become attached to the sacred-self implied in the expectations. Butler argues that performity does more than that: It constitutes the effect it names. Butler (1993) gives the illustration of a judge. When a judge makes a ruling, she or he cites the law. If the judge didn't cite the law, whatever she or he said would not be binding. The power of the judge therefore lies in the citation (iteration/reiteration) of the law. It's the citation that "gives the performative its binding or conferring power" (p. 225). Notice also that something comes into existence through citation. When the judge passes judgment citing the law, the judgment or punishment then exists. Think also of the wedding pronouncement or the act of accepting academic credentials in a graduation ceremony. These declarations are acts of performity: They simultaneously cite an authority and create a state of being. In Butler's scheme, then, heterosexual performity simultaneously cites the norm of hegemonic heterosexuality (thus giving it legitimacy) and produces heterosexual bodies (thus giving it existence). Performity thus has ontological power—it creates the reality in which we live.

Doer in the Deed

I must modify something I just stated. I said, "we are principally embodied in and through sex." According to Butler, that's not quite true. The problem with the statement is it makes it seem like there is a subject ("we") that exists prior to the practices of the body ("are embodied"). To critique this assumption, Butler plays off of Friedrich Nietzsche's idea of the doer and the deed. Most of us feel, talk, and act as if there is a doer behind every deed. In other words, a person acts and exists outside of the practice. In this way, she or he expresses agency—agents act. Nietzsche (1887/1968) turned this idea around when he said, "there is no 'being' behind doing, effecting, becoming; 'the doer' is merely a fiction added to the deed—the deed is everything" (p. 481). In other words, the idea of agency is wrongheaded. The notion that there is an independent person that preexists and is the source of action is not a reflection of the way things actually work. There isn't really a you behind your actions.

The idea that there is a you in back of your actions is, according to Butler (1993), a "presentist conceit, that is, the belief that there is a one who arrives in the world, in discourse, without a history, that this one makes oneself in and through the magic of the name, that language expresses a 'will' or a 'choice' rather than a complex and constitutive history of discourse and power" (p. 228). Here's an illustration. Let's say you are African American. If you had been alive in 1840, it would have been impossible for there to be an African American "you" in back of your actions. African Americans couldn't and didn't exist back then. The only way there can be an American "you" in back of your actions today is that history has socially and culturally produced African Americans. The perception of an African American "you" in back of your actions is, as Nietzsche says, a fiction. It's a conceit that sees the present as somehow less historical than all other time periods that have existed up to this point.

Whatever your subjective sense of you, whether African American or heterosexual, it is historically specific. Most of "you," with all your thoughts and feelings, actually existed prior to your birth. You might be an American, black, white, male, female, a student, a short-order cook, a homeowner, a Visa card user, a licensed driver, an alcoholic, and so on, but most of the meanings and relationships that each of these elements involves existed before you were born. Each identity and subject exists apart from you. *You depend upon them for your intelligibility and viability, not the other way around.* It's thus impossible for "you" to be in back of your actions. Further, what produces a sense of you are your actions: "This repetition is not performed *by* a subject; this repetition is what enables a subject and constitutes the temporal condition for the subject" (Butler, 1993, p. 95, emphasis original).

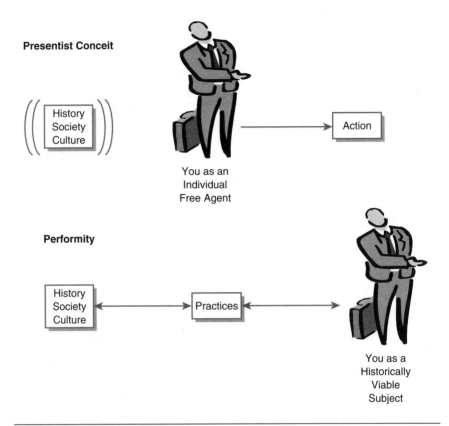

Figure 18.1   Presentist Conceit and Performity

I've pictured this idea in Figure 18.1. There you will see a comparison between presentist conceit and performity. The first set of images, presentist conceit, shows how many people in contemporary societies see themselves. This view of self is powered by the idea of the individual as a free agent, an idea that gained currency as Protestantism, nation-states, and capitalism took hold. As you can see, the individual is aware of history, society, and culture but these are bracketed off and the person is seen as independent of them. Once that is done, we are able to claim the individual as a free agent that acts. Note that this way of looking at things makes the idea of the acting individual the effect of an historically specific ideology.

Butler claims, however, that such a move gives a false sense of self. Society and culture do historically preexist the individual and his or her behaviors. In fact, the only behaviors that are intelligible are those cultural practices that are historically specific. The performity of these practices reiterates society on the one hand and produces the person as a knowing and knowable subject on the other.

One of the things we may not like about the performity model is that it seems to suggest that we're simply robots carrying out the demands of society. But this isn't the case. Notice that all the arrows in this part of the figure are double-headed. This implies that all these sites mutually constitute one another. In other words, under the idea of performity, the individual can influence her or his world. Behaviors, feelings, and ideas don't directly originate with the individual; they are specific to time and place. However, the person is the site through which culture and society are enacted, and every action is in some way particular to the individual. More importantly, as we'll see, the individual may intentionally subvert the practices that cite society. Whether incidental or intentional, the person's actions influence, constitute, and change history, society, and culture.

The main point here is that performity creates the subject, not the other way around. When we act in a way that can be understood by people, we are citing, iterating, and reiterating the textual scripts that have been handed to us. Our intelligibility as embodied persons begins as hegemonic heterosexuality is inscribed upon our bodies through psychodynamic processes, and it continues nonstop through our performative citation of hegemonic heterosexuality.

This citation isn't done by an independent subject that somehow transcends all history; the person's subjectivity, the ability to know and be known, is created through the naming effect of performity. "Sex is, thus, not simply what one has, or a static description of what one is: it will be one of the norms by which the 'one' becomes viable at all, that which qualifies a body for life within the domain of cultural intelligibility" (Butler, 1993, p. 4).

Concepts and Theory: Haunting, Subversion, and Queer Politics

But what about homosexuality? If sex constitutes the body as a meaningful body, and if sex is produced under the hegemony of heterosexuality, then how and where does homosexuality fit in? In order to begin to answer this question, we have to revisit psychodynamic theory for a moment.

As I mentioned earlier, based on Freud, humans begin with only the primitive energy of the id. The id is motivated by the libido, which is psychic power that comes from primary biological urges, particularly sexual energy. These urges come in contact with the human world and thus have to be controlled. Control happens first through the superego. The superego is the voice of society within the individual. It represents the moral restrictions and demands of the external world. The *ego* comes to exist as a mediator between the id and superego. The ego is oriented toward the internal world of the individual and seeks to satisfy the urges of the id within the framework given by the superego. An important way the ego mediates the relationship between the id and superego is through ego defense mechanisms, most notably repression. *Repression* serves to keep the ego unaware of the aggressive urges or painful memories associated with the libido.

According to Butler, this behavior of the ego does two important things. First, it serves to create boundaries between the individual and society and between the individual and her or his instinctual drives. These boundaries represent a crucial step in psychodynamic development—it is because of these boundaries that the individual can become self-aware, reflexive, and experience individual subjectivities. Remember that babies and infants are motivated only by their instinctual drives.

At that stage of development, we are nothing but a series of never-ending desires. During this phase, we also see very little difference between our selves and the world around us. In this sense, we feel ourselves to be a natural part of the environment, with a continuous and unbroken interplay between us and our world.

This condition is very similar to how we think most animals relate to the world, in a nonconscious flow of ongoing stimuli and experience. A prowling leopard, for example, is unaware of itself as a being separate from its natural environment. It instinctually responds to smells and sounds without having to think of itself within the situation. For the leopard, there isn't an awareness of separateness, only of continuous experience. In other words, the leopard doesn't have a self. If the leopard eats, it simply eats. If it has sex, it simply has sex. It isn't *aware of itself* eating—the awareness of self is a separate consciousness apart from the act of eating. The leopard cannot be "embarrassed" if it belches or makes a mess while eating. People, on the other hand, can become embarrassed, because we are aware of ourselves performing the action.

In order to be self-aware, psychodynamic theory argues that humans must first create a boundary between themselves and everything else. This happens principally through the formation of the ego. Through repression, the ego is structurally separated from the body's biological drives (the id) and from the moral demands of society (the superego). By definition, the desires of the id are not aware of themselves. The id is pure desire, just like the drives of the leopard. The demands of the superego initially come from the parent and subsequently society. Again, by definition there is no *self*-awareness present in the superego; it's the voice of the parent that speaks.

The person becomes self-aware when she or he individually negotiates the demands of the id and superego. A woman "finds" herself in the decision to act; her "voice" comes out of the contradictory demands of the id and superego. Interestingly, you and your subjectivity are formed as society denies your fundamental drives and desires, and since society is always politically and culturally constructed, the subjectivity that forms in response is always historically specific and fundamentally political.

In this separateness, there is a sense in which the ego becomes a substitute for the drives of the id. Let's use the eating example again. When my dog Gypsy eats, she is completely engrossed in what she is doing. She is utterly *there*, in the moment, caught up in her own behaviors. It's a different story when I eat. I'm generally aware of how I'm holding my fork; whether or not my mouth is closed while chewing; if I'm eating too fast, too loud, or too much. I'm not completely *there* in the behavior of eating; rather, a large part of me is watching the behavior and judging its appropriateness. In this sense, my ego, my self-awareness, takes the place of the object of my desire, in this case food. This is the second important effect of ego-defense mechanisms, the substitution of the ego for the desire: "The turn from the object to the ego produces the ego, which substitutes for the object lost" (Butler, 1997, p. 168).

These effects are important, but the significant thing here for Butler is how this reflexive turn occurs. Notice what happens to the impulses of the id in psychodynamic theory. They don't go away; they are repressed. There is, then, a sense of loss in the formation of ego. The id, which operates according to the pleasure principle, desires something but can't have it or can't gratify the impulse immediately. The ego develops as the person tries to reasonably satisfy the desires to some degree (the reality principle). As the moral voice of authority (the superego) develops, the ego feels the demands of instinct and of society. In order to negotiate these contrary demands, the ego develops defense mechanisms. While there are several ego defense mechanisms, the most salient here is repression. The individual suppresses or pushes out of her or his consciousness the fundamental yet unsatisfied demands of the id. Therefore, the ego is principally formed and experienced through control and loss.

In Butler's scheme, this loss is experienced as *melancholia*. The "substitution of ego for object does not quite work. The ego is a poor substitute for the lost object, and its failure to substitute in a way that satisfies (that is, to overcome its status *as* a substitution), leads to the ambivalence that distinguishes melancholia" (Butler, 1997, p. 169, emphasis original). The problem with this kind of loss is that it is at once constitutive yet unspeakable. In other words, our self (ego) is created or constituted through loss; however, this loss not only occurs before it could be articulated, during early childhood, but the loss is also repressed and thus part of the unconscious—the loss creates the ego, and at the same time the ego denies the loss.

But because of the nature of loss, it haunts us. There is a sense of loss, but just what this loss is never fully appears. What we are left with is an ambivalent kind of sadness or melancholy. According to Butler (1997), this "melancholy is precisely what interiorizes the psyche" (p. 170) and permeates "the body with a pain that culminates in the projection of a surface, that is, a sexed morphology" ((1993, p. 65). Both the self and the body are inscribed and circumscribed through loss. But what is this fundamental loss?

As we've seen, Butler argues that the body and ego are formed under historical conditions. For us, these conditions are set by the hegemonic norm of heterosexuality. The loss, then, is the refusal of the possibility of sexual or sensual connections between members of the same "sex." Remember that the **materialization** of the body, the body's ability to exist as a physical body, is normative. Biological sex is regulated socially. That's the underlying reflexive idea behind the title of Butler's 1993 book, *Bodies That Matter: On the Discursive Limits of "Sex."* Clearly, bodies matter in the sense that they are important, but more significantly, bodies *are* matter: They have material substance to them. Yet the material is socially regulated and created. The physical nature of the body is only intelligible to us under the dominion of a historically, culturally, and politically specific understanding of heterosexuality. Thus, these bodies we have are conditioned and given material through performity, the inscribing and citing of heterosexuality. This performity of the hegemonic norm, along with the Oedipal law, constitutes the modality of the materialization of the body—the method or channel through which the body becomes a body.

Further, the ego or our subjective sense of self is formed under these same conditions. By its very nature, the ego is denial. It is formed in order to control and refuse the impulses of the id. The single most important energy coming from the libido is sexual. In this era, our claim to a "biological" sex is the basic way we understand our body. It's more fundamental than race or weight or any other "physical" characteristic. And the single most important denial or loss that occurs under the hegemonic norm of heterosexuality is, by definition, the lost possibility of homosexuality.

all texts are inherently based on contradictions; discourse assumes that people are brought under political and social power through their own use of discourse (they thus become subjects of the discourse), yet discourse itself is subject to abrupt breaks and discontinuities; and media reproduction assumes that the use of media destroys information and the social text. Given these ideas, we can say that in postindustrial society, human reality is at best fragmented and free-floating and at worst filled with simulations of realities that never existed. Further, if this is true about human reality, then the subject is dead or simply a terminal upon which media images play themselves out.

But perhaps you don't feel that these conclusions adequately represent your world and reality. Maybe you feel that your world is rather substantial and that your identities and sense of self are centered and cohesive. However, your feeling these things isn't enough for disciplined or scholarly thought. How would you theoretically critique these ideas? One way to go is to determine what these perspectives leave out. If we compare and contrast the focuses and assumptions of the different perspectives, as in the table I'm suggesting you make, then we can see that postmodernism and poststructuralism are focused on language, culture, and text, but leave out people and their situations. So what would theories that focus on the situation or individual bring to this discussion? What do they tell us about human reality and meaning? (Notice that we know where to look because we've identified what the perspectives leave out.)

Briefly, those theories that focus on the situation tell us that meaning emerges through ongoing interactions (symbolic interaction), that the self is the central organizing feature of the social encounter (dramaturgy), that people achieve social order situationally (ethnomethods), that social exchange results in diffuse and long-term social obligations (exchange theory), and that situations are chained together through emotional energy and cultural capital, the very things that cause interactions to occur in the first place (ritual theory).

Those theories that consider the individual tell us that people are motivated in interactions to produce ontological security (structuration), that they receive primary socialization that sets their reality (social constructivism), that they tend to develop significant ego-identities (dramaturgy) through emotionally dynamic interactions (ritual theory), and that the very body of the individual is inscribed (constructivist structuralism and queer theory).

These theoretical ideas indicate that there are factors at the level of the situation and the individual that tend to produce a strong sense of meaning, reality, and self. So, are postmodernism and poststructuralism wrong? Nope. Theories aren't ever "right" or "wrong." They can be more or less insightful, logical, or powerful, but they can't be wrong in an absolute sense (this is one reason why I said in the introduction that I don't believe any of these theories—none of them are right, but neither are any of them wrong).

So, what can we say in conclusion? Most importantly, we've pointed out a gap in both the textual and situational approaches—textual theories typically leave out the encounter and situational theories usually neglect broader cultural issues. In general, then, we can say that poststructuralism and postmodernism appear to make conclusions about the individual (the subject is dead) without adequately taking into account either the individual or her or his situation. Notice the word *appear:* We can't say anything definite at this time—much more research and thinking is needed. But we have identified a couple of major issues to look for in our research.

First, it appears that a major issue is "the situation." If social order, meaning, self, and identity are all produced in situations that are connected by daily rounds (dramaturgy), diffuse reciprocal obligations (exchange), and varying degrees of emotional intensity (ritual), then poststructuralism and postmodernism don't have much to offer us. However, has anything happened to "the situation" in the recent past? Yes! Structuration theory indicates that the situation has been stretched out through time–space distanciation. And, postmodern theory implies that "the situation" is no longer limited to face-to-face encounters: Situations are increasingly presented, organized, and interpreted through the mass media, and there are new virtual situations. In addition, situations don't occur in a vacuum, they occur in cultural (not to mention structural) contexts, which brings us back to our textual theories as well.

A second major issue that we've identified concerns what we might call the internalization process: the transition point between the individual and the social (would it be helpful to bring in the idea of boundaries from systems theory?). Most of our individual theories assume a rather nonproblematic process that results in habitus, ontological security, and objective facticity. But what all goes into the socialization process? Culture/text is certainly a major element. For example, queer theory tells us that our bodies are inscribed with hegemonic heterosexuality. And there are effects: This particular kind of embodiment has resulted in haunting. So what happens when the internalization process occurs under conditions of postmodern culture?

Notice what we've done: We've used some of the major issues associated with theoretical perspectives to identify questions and areas of research. This kind of contrast and comparison takes place at the categorical level, which is one reason why understanding theoretical perspectives can be such a powerful tool. However, the true power of such an approach is based on depth as well as breadth. Since we started in this book with depth, the exercise I'm asking you to do—generalizing the particulars of a theory to its perspective—will assure that you have breadth as well.

Because theory isn't just something you know, it's something you do, I also want you to notice *how* we did what we did. *We opened up a theoretical space by simply contrasting and comparing ideas.* I've asked you to do this throughout the book precisely because it is so basic to what we do as theoretical sociologists—it's one way we form theories. We can see this method of critique in the work of Pierre Bourdieu, Michel Foucault, Jacques Derrida, and Jean Baudrillard. Once this space has been opened, we can fill it through synthesizing theories. *Theoretical synthesis is a work of imagination.* In it we create a new theory by arranging elements from different theories in a creative way. In this book, we can see this work explicitly with Randall Collins, Janet Chafetz, Anthony Giddens, and Judith Butler.

Pitting ideas and concepts against one another and then synthesizing elements is a well-worn path of theorizing, as you can see from the examples I've given. However, what do you think Dorothy Smith would say about this method of theorizing? Smith would say that beginning in theory to produce theory runs the

Robert Young's *Postcolonialism: A Very Short Introduction.* Above all else, it will tell you why you should care. Next, Said's *Orientalism* and Fanon's *Black Skins, White Masks* and *The Wretched of the Earth* should all be read. I then suggest that you follow up on the further readings that Young recommends in his book. It will remake your mind.

But I have another reason for mentioning postcolonialism. I said I was using it as an example of the idea of perspectives in contemporary theory, and that's partially true. My intent is a bit more significant, however. Earlier we entertained the idea that our ability to have a significant thought is an interesting theoretical question and not a personal issue. What we think and how we think is clearly influenced by the cultural milieu in which we live. So, for example, thinking in a postmodern society today is a different kind enterprise from thinking in the traditional society of pre-Christian Jerusalem.

However, there is one thing that strikes me as I read even the most postmodern or poststructuralist of authors: Every one of them assumes her or his reader to be capable of logical, critical, and reflexive thought. That our subjective existence is linked to the discourse of the age is beyond doubt. But within that existence, even within the mind-numbing world of postmodernism, we can have a thought. We can become aware and, because we are symbol-using creatures, we can take our thought a step further than its taken-for-granted base.

Thus, in bringing postcolonialism into our discussion, I want to emphasize a central point: *To have a thought is to be post.* I don't mean this statement as a catchphrase or truism. If we accept the implications of discourse, then having a thought as compared to being handed a thought always involves moving from a position of taken-for-grantedness. To even be aware of our discourse implies some movement out of it. How far we move depends on the distance between symbolic spaces. Though I admit the limitations of analytic dichotomies, the basic distinction in symbolic space is between within and without. When we move outside our symbolic space in the West, it is usually in reference to such collective distinctions as gender, sexuality, race, ethnicity, class, religion, and so forth. Postcolonialism, however, asks us to take a larger step—a step from one world into another, from everything most Westerners believe about modernity, progress, technology, democracy, and capitalism to worlds that are defined and experienced in terms of diasporas, periphery, imperialism, dislocation, refugeeism, and terrorism.

What contemporary theory and this book are asking us to do is to move. In this sense, modernism is larger than an historical period or specific social factors—it is a perspective of the mind and an attitude of the heart. To be modern, then, is to uncritically believe in the hope of technology and government; it is to live within the confines of a materialized body, to accept a simplistic and political reduction of space and time, and to keep oneself within the confines of identities of certainty.

In addition, one of the things that post-thinking implies is that it is impossible to move backward or forward. To move backward implies trying to think in terms of cultures past, like trying to imagine what it was like to be a Hopi before the Europeans came. No matter how close we may think we get, whatever symbolic space we create is always and ever a "traditional-Hopi-as-imagined-by-a-21st-century-person." The same is true about seemingly closer subjectivities, such as the idealized American family of the 1950s. So we can't truly move backward in our thought and we can't move forward either. Moving "forward" exists only as the result of an ethical standard, and currently that standard is some form of the modern ideal of progress.

To have a thought, then, is to engage in post-thinking. It isn't a call to a bygone era, nor is it valued as a step forward; it is simply a step outside. Since today the most significant "inside" is the symbolic space wrapped around the idea of modernity, to have a thought is to be postmodern: to think outside the confines of modernist values and ideas. To have a thought is thus to be liminal—it is be outside but not yet arriving. In this way, Nietzsche misplaced his thinker. Rather than "in the middle of the road," perhaps we should see a person having a thought as being on a staircase—neither going up nor down but certainly situated between, in a conceptual space that is equally purposeful and unnerving (see Allan, 2005c).

Contemporary theory is vibrant and unruly precisely because it is post in this fashion. Theory today is also the most exciting it has been since the time of Marx, Durkheim, and Weber, where social theory was a main thread in public discourse and media. The book you hold in your hands invites you into this turbulent sea of ideas. Contemporary theory is an invitation to rethink society and the individual, power and possibilities, cultures and realities, time and space, human bodies and emotions, the earth and the risks we produce, and the ethics of it all. Contemporary theory is an invitation to move from our center and to see through different perspectives. How far are we willing to move, to be unsettled? What kinds of responsibilities are we willing to accept as a result of seeing differently? What will we make of our world? These are the questions of contemporary theory.

References

Alexander, J. C. (1985). Introduction. In J. C. Alexander (Ed.), *Neofunctionalism*. Beverly Hills, CA: Sage.

Allan, K. (2005a). *Explorations in classical sociological theory: Seeing the social world*. Thousand Oaks, CA: Pine Forge Press.

Allan, K. (2005b). *Where is reality?* Manuscript submitted for publication, University of North Carolina, Greensboro.

Allan, K. (2005c). *Post with a purpose*. Manuscript submitted for publication, University of North Carolina, Greensboro.

Allan, K., & Turner, J. H. (2000). A formalization of postmodern theory. *Sociological Perspectives, 43*(3), 363–385.

Al Qaeda Training Manual. Retrieved October 27, 2004, from http://www.usdoj.gov/ag/trainingmanual.htm

Anderson, P. (1998). *The origins of postmodernity*. London: Verso.

Bandura, A. (1977). *Social learning theory*. New York: General Learning Press.

Barthes, R. (1967). *Elements of semiology* (A. Lavers & C. Smith, Trans.). London: Jonathan Cape. (Original work published 1964)

Baudrillard, J. (1981). *For a critique of the political economy of the sign*. St. Louis, MO: Telos Press. (Original work published 1972)

Baudrillard, J. (1975). *The mirror of production* (M. Poster, Trans.). St. Louis, MO: Telos Press. (Original work published 1973)

Baudrillard, J. (1987). When Bataille attacked the metaphysical principle of economy. *Canadian Journal of Political and Social Theory, 11*, 57–62.

Baudrillard, J. (1993a). *Baudrillard live: Selected interviews* (M. Gane, Ed.). New York: Routledge.

Baudrillard, J. (1993b). *Symbolic exchange and death* (I. H. Grant, Trans.). Newbury Park, CA: Sage. (Original work published 1976)

Baudrillard, J. (1994). *Simulacra and simulation* (S. F. Blaser, Trans.). Ann Arbor: The University of Michigan Press. (Original work published 1981)

Baudrillard, J. (1995). *The Gulf War did not take place* (P. Patton, Trans.). Bloomington: Indiana University Press.

Baudrillard, J. (1998). *The consumer society: Myths and structures* (C. Turner, Trans.). London: Sage. (Original work published 1970)

Bauman, Z. (1992). *Intimations of postmodernity*. New York: Routledge.

Bem, S. L. (1974). The measurement of psychological androgyny. *Journal of Consulting and Clinical Psychology, 42*, 155–162.

Bentham, J. (1996). *An introduction to the principles of morals and legislation* (J. H. Burns & H. L. A. Hart, Eds.). New York: Oxford University Press. (Original work published 1789)

Berger, P. L. (1967). *The sacred canopy: Elements of a sociological theory of religion.* New York: Anchor.

Berger, P. L., & Luckmann, T. (1966). *The social construction of reality: A treatise in the sociology of knowledge.* New York: Anchor.

Bernauer, J. W., & Mahon, M. (1994). The ethics of Michel Foucault. In G. Gutting (Ed.), *The Cambridge companion to Foucault.* Cambridge, UK: Cambridge University Press.

Blau, P. M. (1968). Social exchange. In David L. Sills (Ed.), *International encyclopedia of the social sciences.* New York: Macmillan.

Blau, P. M. (1994). *Structural contexts of opportunities.* Chicago: University of Chicago Press.

Blau, P. M. (1995). A circuitous path to macrostructural theory. *Annual Review of Sociology, 21,* 1–19.

Blau, P. M. (2002). Macrostructural theory. In Jonathan H. Turner (Ed.), *Handbook of sociological theory.* New York: Kluwer Academic/Plenum.

Blau, P. M. (2003). *Exchange and power in social life.* New Brunswick, NJ: Transaction.

Blau, P. M., & Meyer, M. W. (1987). *Bureaucracy in modern society* (3rd ed.). New York: McGraw-Hill.

Blumer, H. (1969). *Symbolic interactionism: Perspective and method.* Berkeley: University of California Press.

Blumer, H. (1990). *Industrialization as an agent of social change: A critical analysis.* Chicago: Aldine.

Bourdieu, P. (1984). *Distinction: A social critique of the judgment of taste* (R. Nice, Trans.). Cambridge, MA: Harvard University Press. (Original work published 1979)

Bourdieu, P. (1985). The genesis of the concepts of *habitus* and of *field. Sociocriticism, 2,* 11–29.

Bourdieu, P. (1989). Social space and symbolic power. *Sociological Theory, 7*(1), 14–25.

Bourdieu, P. (1990). *The logic of practice* (R. Nice, Trans.). Stanford, CA: Stanford University Press. (Original work published 1980)

Bourdieu, P. (1991). *Language and symbolic power* (J. B. Thompson, Ed.; G. Raymond & M. Adamson, Trans.). Cambridge, MA: Harvard University Press.

Bourdieu, P. (1993). *Outline of a theory of practice* (R. Nice, Trans.). Cambridge, UK: Cambridge University Press. (Original work published 1972)

Bourdieu, P., & Wacquant, L. J. D. (1992). *An invitation to reflexive sociology.* Chicago: University of Chicago Press.

Boyne, R. (1990). *Foucault and Derrida: The other side of reason.* London: Routledge.

Braithwaite, R. B. (1953). *Scientific explanation: A study of the function of theory, probability and law in science. Based upon the Tarner lectures, 1946.* Cambridge, UK: Cambridge University Press.

Brown, D. (2001). *Bury my heart at Wounded Knee: An Indian history of the American West* (30th anniversary ed.). New York: Owl Books.

Business and Professional Women's Foundation. (2004). *101 facts on the status of women.* Retrieved May 9, 2005, from http://www.bpwusa.org/i4a/pages/index.cfm?pageid=3301

Butler, J. (1993). *Bodies that matter: On the discursive limits of "sex."* New York: Routledge.

Butler, J. (1997). *The psychic life of power: Theories in subjection.* Stanford, CA: Stanford University Press.

Calhoun, C. (2003). Pierre Bourdieu. In G. Ritzer (Ed.), *The Blackwell companion to major contemporary social theorists.* Malden, MA: Blackwell.

Cassirer, E. (1944). *An essay on man.* New Haven, CT: Yale University Press.

Chafetz, J. S. (1990). *Gender equity: An integrated theory of stability and change.* Newbury Park, CA: Sage.

Charon, J. M. (2001). *Symbolic interactionism: An introduction, an interpretation, an integration* (7th ed.). Upper Saddle River, NJ: Prentice Hall.

Chodorow, N. (1978). *The reproduction of mothering: Psychoanalysis and the sociology of gender.* Berkeley: University of California Press.

CNN (2003). *Commander in Chief lands on USS Lincoln.* Retrieved July 4, 2005, from http://www.cnn.com/2003/ALLPOLITICS/05/01/bush.carrier.landing/

Cohen, A., Baumohl, B., Buia, C., Roston, E., Ressner, J., & Thompson, M. (2001, January 8). This time it's different. *Time, 157*(1), 18–22.

Collins, P. H. (2000). *Black feminist thought: Knowledge, consciousness, and the politics of empowerment* (2nd ed.). New York: Routledge.

Collins, R. (1975). *Conflict sociology.* New York: Academic Press.

Collins, R. (1986). Is 1980s sociology in the doldrums? *American Journal of Sociology, 91,* 1336–1355.

Collins, R. (1987). Interaction ritual chains, power and property: The micro-macro connection as an empirically based theoretical problem. In J. C. Alexander, B. Giesen, R. Münch, & N. J. Smelser (Eds.), *The micro-macro link.* Berkeley: University of California Press.

Collins, R. (1988). *Theoretical sociology.* San Diego, CA: Harcourt Brace Jovanovich.

Collins, R. (1993). Emotional energy as the common denominator of rational action. *Rationality and society, 5:* 203–230.

Collins, R. (1998). *The sociology of philosophies: A global theory of intellectual change.* Cambridge, MA: Belknap Press.

Collins, R. (2004a). *Interaction ritual chains.* Princeton, NJ: Princeton University Press.

Collins, R. (2004b). Rituals of solidarity and security in the wake of terrorist attack. *Sociological Theory, 22,* 53–87.

Condon, W. S., & Ogston, W. D. (1971). Speech and body motion synchrony of the speaker-hearer. In D. D. Horton & J. J. Jenkins (Eds.), *Perception of language.* Columbus, OH: Merrill.

Cooley, C. H. (1998). *On self and social organization* (H. Schubert, Ed.). Chicago: University of Chicago Press.

Coser, L. (1956). *The functions of social conflict.* Glencoe, IL: The Free Press.

Dahrendorf, R. (1959). *Class and class conflict in industrial society.* Stanford, CA: Stanford University Press. (Original work published 1957)

Danto, A. C. (1990). The hyper-intellectual. *New Republic, 203*(11–12), 44–48.

Davidson, A. I. (1994). Ethics as ascetics: Foucault, the history of ethics, and ancient thought. In G. Gutting (Ed.), *The Cambridge companion to Foucault.* Cambridge, UK: Cambridge University Press.

Davis, K., & Moore, W. (1945). Some principles of stratification. *American Sociological Review, 10,* 242–247.

Derrida, J. (1978). *Writing and difference* (A. Bass, Trans.). Chicago: University of Chicago Press. (Original work published 1967)

Derrida, J. (1997). *Of gramatology* (G. C. Spivak, Trans.). Baltimore: The Johns Hopkins Press. (Original work published 1967)

Derrida, J. (2001). "I have a taste for the secret." In G. Donis & D. Webb (Eds.), *A taste for the secret: Jacques Derrida and Maurizio Ferraris* (G. Donis, Trans.). Cambridge, UK: Polity Press. (Original work published 1997)

Drake, J. (1997). Review essay: Third Wave feminisms. *Feminist Studies, 23*(1), 97–108.